FLOOR, DIFFUSER AND GROUND EFFECT

It is the first book in a large and special series of books, dedicated to motorsport in general; it will cover aerodynamics, suspension, engines, dynamics, etc. ..... Everything you need to learn how to design a full car.

The aim of this series is also to say that I would like to teach again in a university.

I hope that this series will be a success and that I will be able to transmit all my knowledge and all my experience.

@TimoteoBriet

Floor & Diffuser

FLOOR

There is a phenomenon that exploits the existence of a flat surface under the car to generate downforce. A car's floor can be designed in a way that with minimal resistance it can generate downforce; the trick is the ride height of distance between the floor and the track.

This part generates the largest amount of downforce in the car:

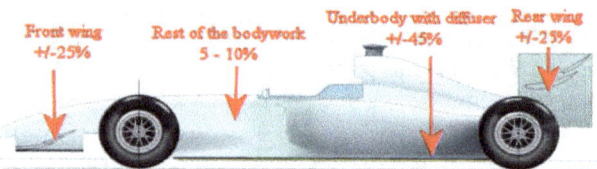

On the internet we can find many quantifications of the amount of downforce that it is capable of generating, in relation to other elements of the car; in any case this load will be very big relative to the others; hence we should take it seriously the optimization of this device.

Take a flat surface close to the ground; a flow travels beneath it; Using Bernoulli we can see that by reducing the section the velocity will increase and pressure will drop sucking air up; this is the essence of the ground effect:

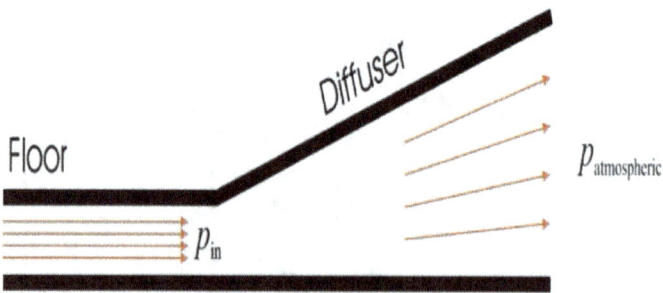

It looks simple at first and in reality it is, but precisely because it is so simple it has problems and variables to consider: it is very sensitive to initial conditions. But in the diffusor, the section is big, so by Bernouilli, the pressure is big iiii (contradiction ??) that we will see after (diffuser).

Suppose now that when air is traveling under the floor, we introduce air into the depression created; it is obvious to say that downforce decreases, since depression will not be as big as previously:

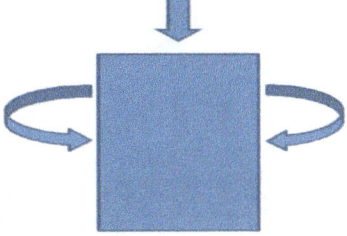

In order to increase the depression in the ground, the car tilts (higher at the rear than at the front); this angle is called "pitch"; it can vary greatly from car to car, but barely exceeds 2º:

➔ Red Bull – Adrian Newey rake:

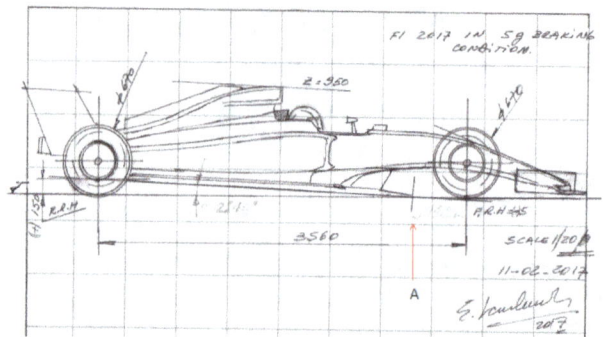

IN 2016 season, Adrian Newey introduce a rake very big. That is:

The point "A" is very behind, in order to allow go down the front wing. This system, allow have a lot down force in front wing and big downforce in rear wing also (in corner….). If the speed decrease, the downforce in front decrease and also in rear wing (low incidence angle….). The same than if the speed increase. Beside, is possible to use springs softer, so good grip….

This is a amazing idea…. DOUBLE DRS….

Ferrari in 2017 season, think the same:

To prevent the air that is right on the edge of the low pressure area be introduced into the depression, some barriers are placed; they are called side skirts. The floor turns into a "pipe", where it would be possible to apply Bernoulli....

Lotus T-78 and T79:

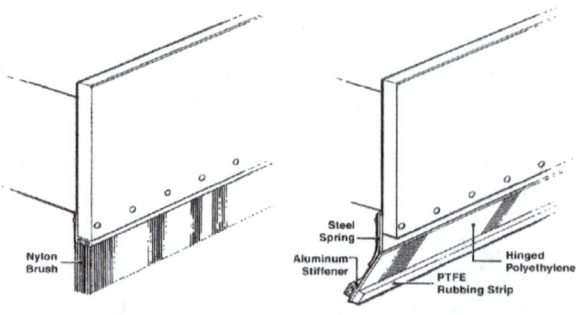

Nylon
Brush

Steel
Spring

Aluminum
Stiffener

Hinged
Polyethylene

PTFE
Rubbing Strip

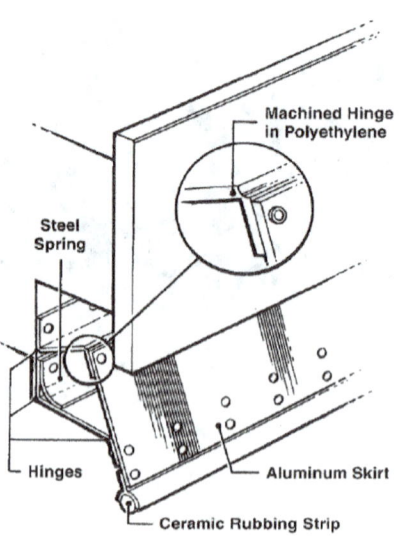

Machined Hinge
in Polyethylene

Steel
Spring

Hinges

Aluminum Skirt

Ceramic Rubbing Strip

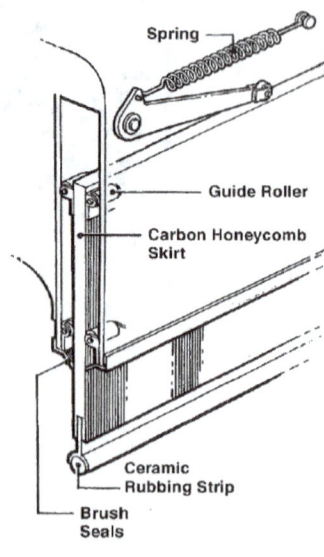

Spring

Guide Roller

Carbon Honeycomb
Skirt

Ceramic
Rubbing Strip

Brush
Seals

Without taking into account risk issues (any variation of the ride height produces a large variation of downforce and therefore grip), these side skirts are an extraordinary idea. The danger lays in the fact that if for any reason the car stopped having these skirts "activated" a lot of downforce is automatically lost, the car would "take off". Imagine the analogy of a compressed spring when we suddenly release it.

Suppose we could design the floor of the car so that it was not flat; i.e. a great way to generate more depression (two designs):

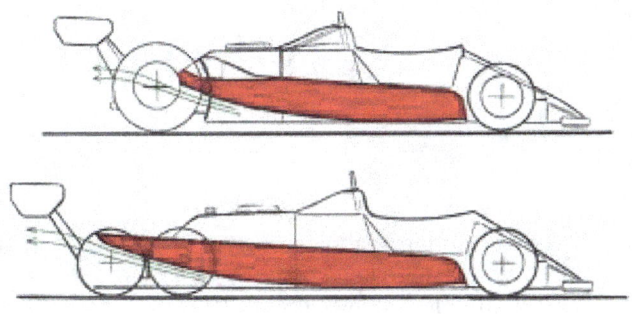

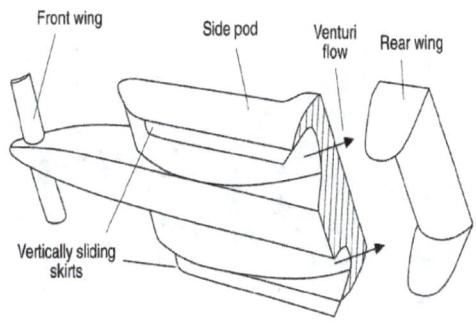

This allows the floor to generate a huge amount of downforce; but for safety, the car's floor is only allowed to be completely flat and wood planks must be placed to know whether we are too close to the ground.

Images Joseph Katz:

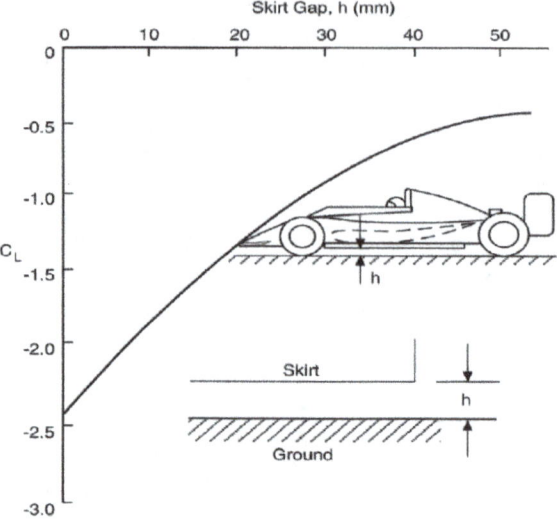

Skirt Gap, h (mm)

By Normative technical in F1 for example, in the car floor must exist a wood plate, in order to allow more security (it is dangerous any height to track) and validate the height:

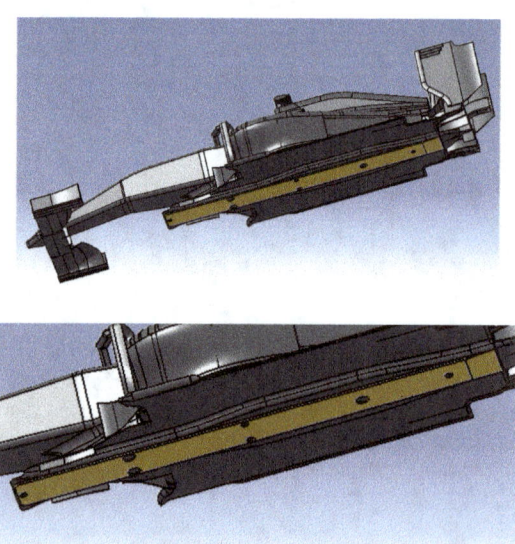

However, obviating the rules, it is not possible to place the floor at any distance to the ground:
Suppose the following floor and diffuser design:
The leading edge radius is 10 mm (start floor):

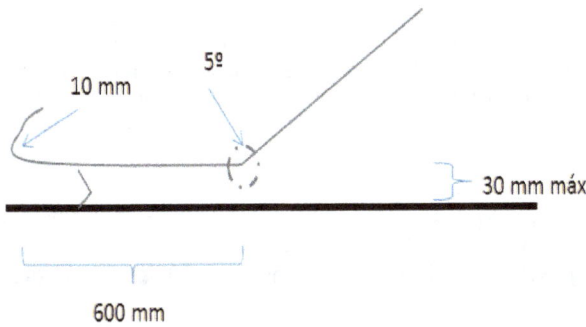

Colin Chapman studied this floor concept obtaining two important conclusions:

- The change in angle had to be done with a edge radius of 300mm (circle discontinuous):
- The relationship between the inlet area and the outlet area could not exceed 1/22.

Chapman introduced this unoptimized concept study that you can see in the image above; the car was unmanageable; when speed increased the generated load closed the gap between the floor and the ground, but when he did the inlet-outlet relationship grew above 22, which formed a plug at the inlet thus not all air could enter below the car; this entailed a reduction in the load and the car started bouncing up and down. That is very important, because this relation between input-output, is essential in order to have a good difusser.

This is known as "purpose"; to eliminate this dynamic issue, 2 improvements are introduced and discussed. Studying this phenomenon means to study a dynamic phenomenon which is complex by definition. This is an event that is embodied in racing cars with great variation of pitch at high speeds; this forces the driver to release throttle. Many tests have shown that oscillations can be up to 30 mm around frequencies of 5-6 Hz.

This leads to accelerations exceeding 1 g. The rear springs displacement are progressively increased until the driver feels the danger and releases the throttle.

The floor must have a uniform pressure distribution; this means that it works perfectly generating downforce.

Let's take a look at the areas that generate most of the downforce of the car; as we can see on the image below wings and floor are the ones that generate the most:

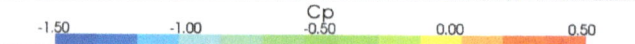

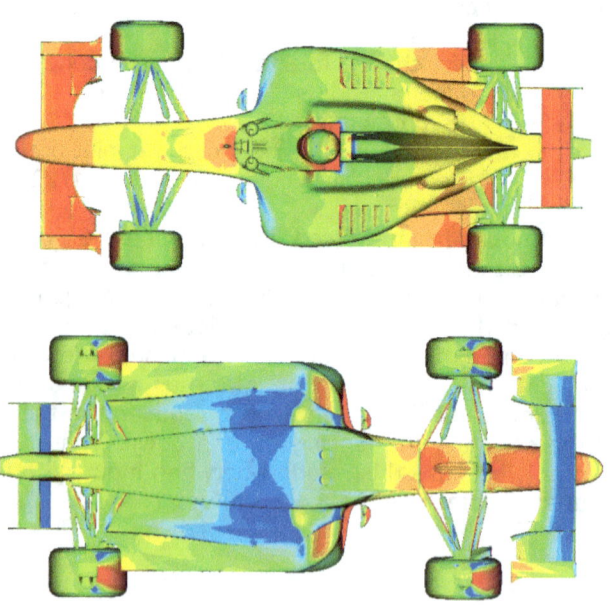

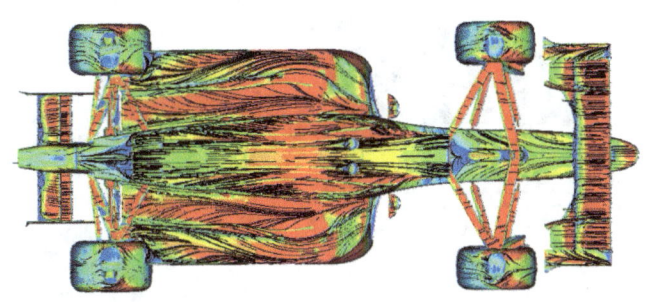

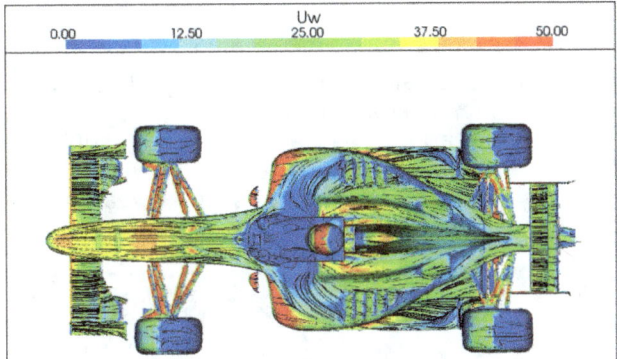

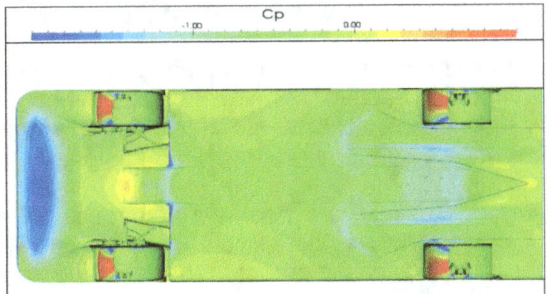

Another think very important, for designing a difusser or floor, is the boundary layer creation: this boundary layer, reduce the section in floor; may be also, to create a stop by viscosity:

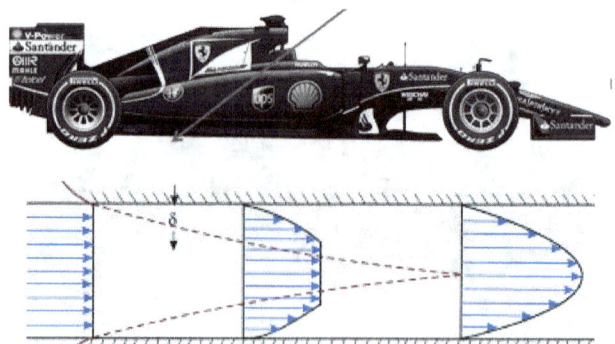

We have 2 other ways to generate downforce on the floor:

- Suck air from the bottom with an exhaust-fan. Initially this fan was said to cool the gearbox as a way of justifying its installation on the car. However, its real objectives where soon revealed to be different...

As the fan was directly linked to the engine, in a straight line it reached many rpm, generating lots of useless downforce; therefore the blade's angle was varied to reduce suction and drag.

- To seal the floor aerodynamically, similar to how side skirts function, but without physical elements. The method tries to generate a high energy vortex moving longitudinally along the car as a barrier that tries to prevent the circulation of air from the area of high pressure to the area of low pressure: the principal Vortex longitudinal, is named Vortex Y250:

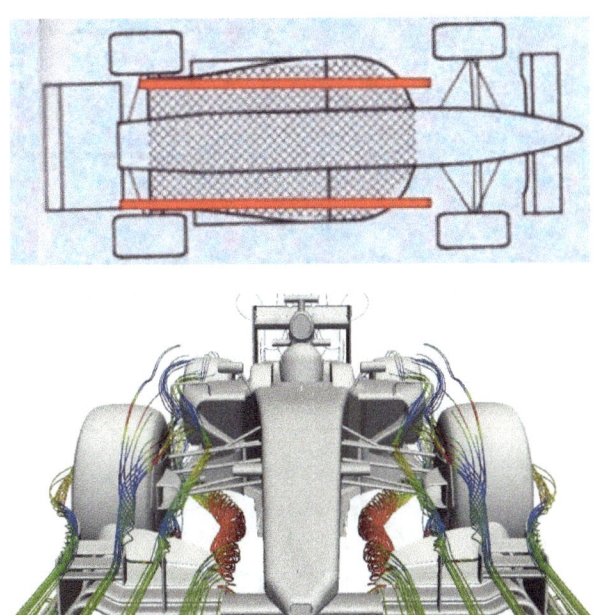

These vortices can be generated using devices placed on:

- The front wing.
- The floor; this is the case of the Le Mans series. Its geometry is the following (the hypotenuse is the one that is to contact with the car's floor).
- Etc….

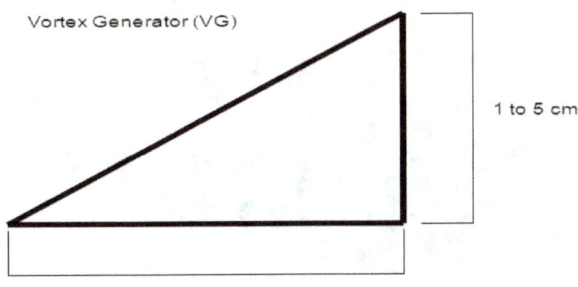

Vortex Generator (VG)

1 to 5 cm

2 to 10 cm

Also:

Wind

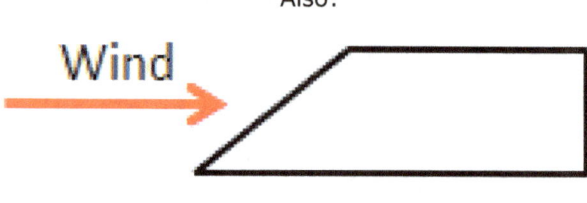

Air Flow.

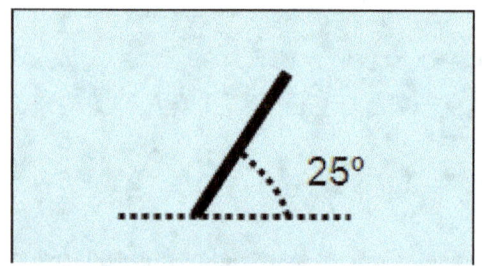

25°

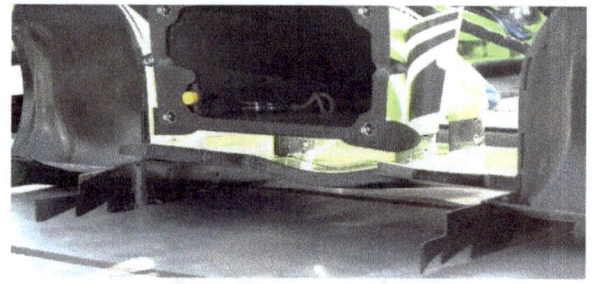

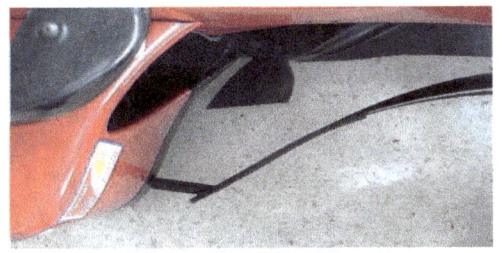

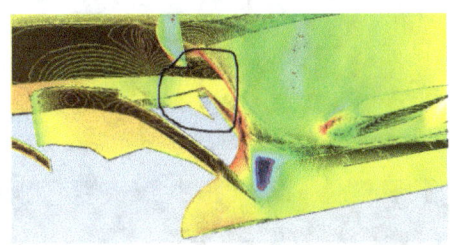

The next images, we can see one system in order to add energy to vortex Y250:

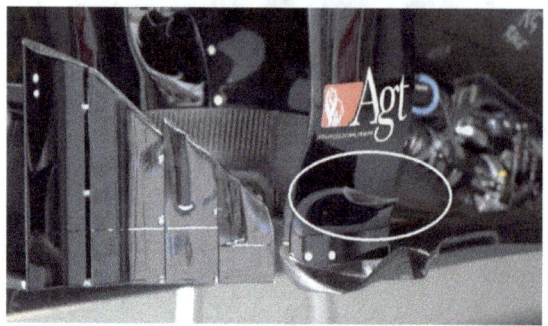

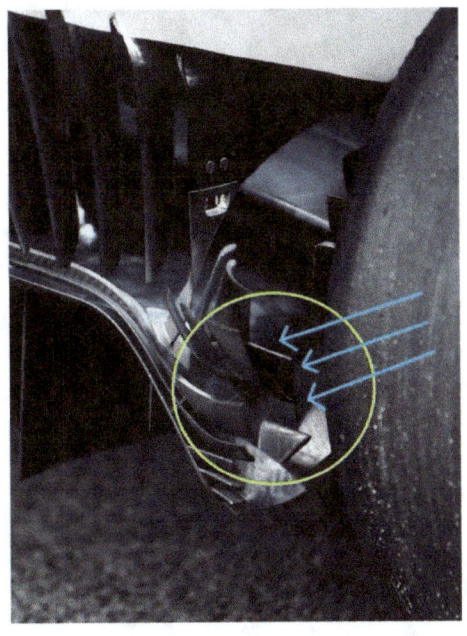

We can produce similar as before, marginal vortexes longitudinally along the outside of the floor.

There are laterals vortices created by the front wing (one type already see it); the complexity of the design is due to the extreme sensitivity of the flow generated, depending on its geometry:

These vortices either:

- Act as a barrier like the method above.
- Act removing air from the area of low

pressure and surroundings, rotating the flow in a certain direction.

The direction of rotation of the vortices should be counterclockwise, to remove air and prevent the introduction of more air under the car.

It is very important to observe the direction and path of the generated vortices (always from high pressure to low pressure):

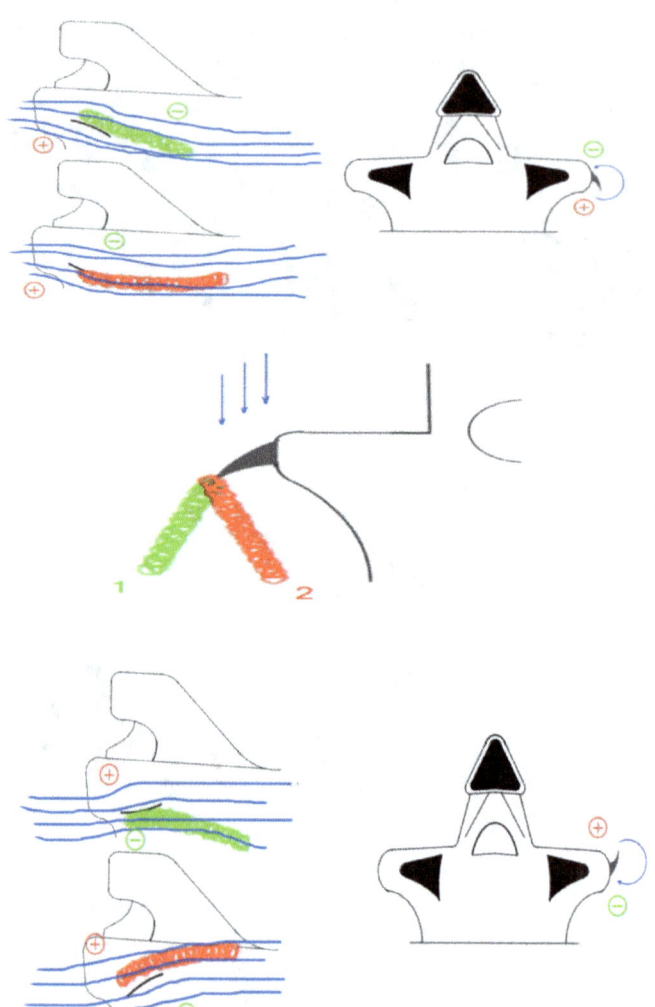

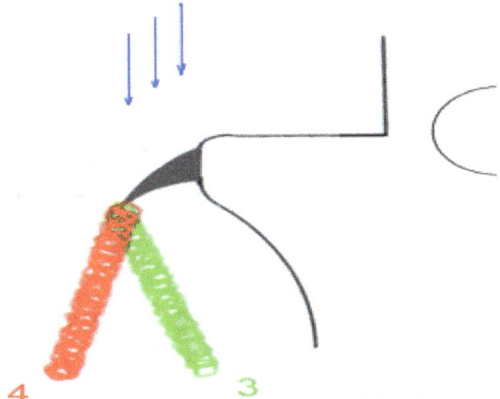

As we know air goes from the high pressure area to the low pressure area, we can design an element to rotate air as we need it to do.

It is very important to know whether the vortex tends to fall or raise:

	Incidence	Vertical	Rotation
1	+	falls	counterclock wise
2	+	raises	clockwise
3	-	falls	clockwise
4	-	raises	counterclock wise

One think very important to know, is the possibility of creation of vortex, from another vortex; that is:

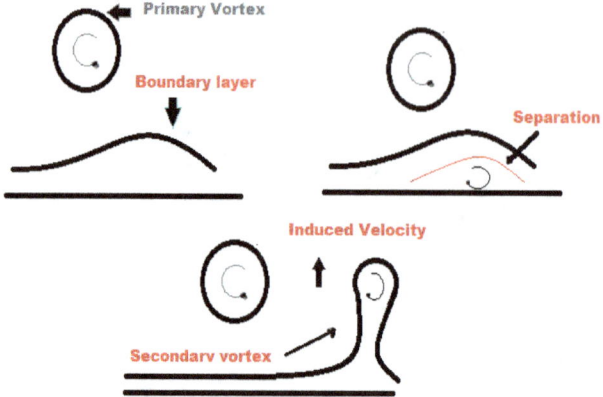

And may be produce vortex, without want its:

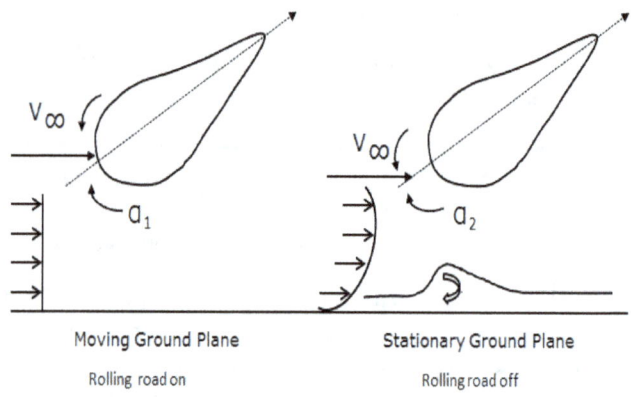

Moving Ground Plane

Rolling road on

Stationary Ground Plane

Rolling road off

Knowing how the vortex rotates involves knowing whether we are "extracting" or not air in a given area.

Imagine a sidepod seen from the front: we need the vortex to rotate in the direction marked by the red line. In this way we are "ejecting" all the air that wants to enter the floor of the car, disturbing and spoiling the downforce generated by the floor:

The cambered barge board (difference pressure), produce a vortex very important (as a skirt in ground effect).

These vortexes are very important: increase the ground effect, improve the refrigeration, improve the diffuser function and reduce drag in rear wheels.

We can now take a look at how F1 teams like Red Bull, Mercedes and Ferrari do this, analyzing the quality of the flow that runs on both sides of the car.

1) Red Bull controls vortexes for almost the entire side of the car and up to the rear wheel, where the vortex travels on the floor and moves to the upper part of the diffuser attracted by the exhaust flow that travels between the "end fence" of the

diffuser and the inner side of the rear tire.

The rear wheel's pressure makes another flow travel inward below the vortex with more energy, entering the bottom part of the diffuser creating a blow of air increasing the air's energy increasing downforce.

2) In the second picture we can see that the vortex does not flow parallel to the side of the car and it opens impacting on the outside of the rear wheel, which can create aquaplaning and less downforce, so they should have a lower rear height clearance losing "rake"; i.e. generating less downforce.

3) In the Ferrari, the problem is the same or worse than the Mercedes. Very far from the quality control of lateral vortex produced by the Red Bull.

Let's take a look at a series of images where we can see this beautiful Vortex Y250: Amazing images:

Today practically all teams, include devices in bargeboards (named hammer heads):

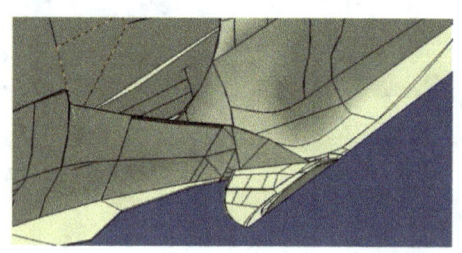

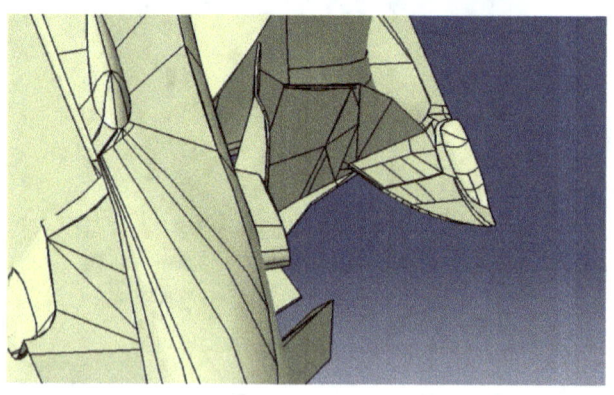

These bargeboards are responsible for diverting UP (red arrow) the air that collides with the sidepod:

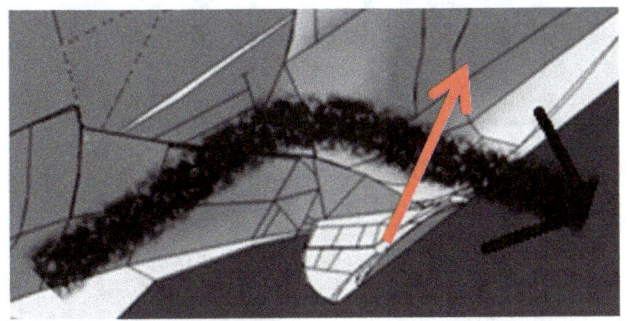

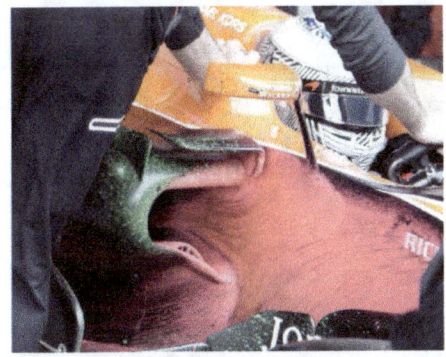

The same the down edge in ground, for diverting UP the air from front inlet:

The issue here is to remove the air from the floor's depression area.

Another place where you can place the vortex generators is on the floor of some motorsport series; the car may have a number of channels where the elements in question can be installed:

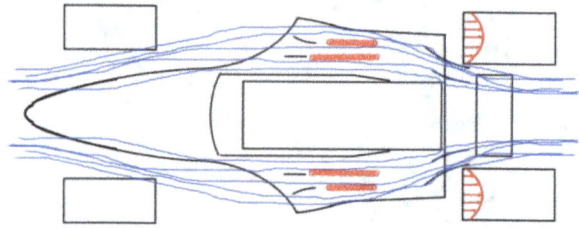

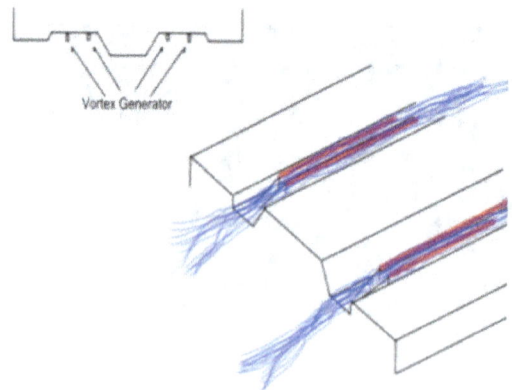

Pillar vortex:

This is one of the most typical vortices because of how it is generated and its transcendence.

These vortices are formed whenever the flow travels across right angles. These vortices produce not only drag, but also causes the aerodynamic devices to malfunction downstream. Anyway, we can generate these vortices on purpose with the intention of using them: one of the objectives could be stability:

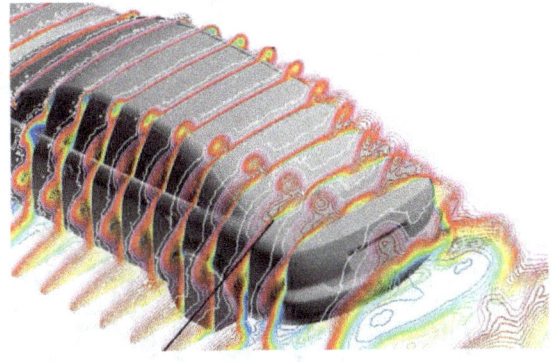

In the body called Ahmed (used to validate CFD models), it is desirable that these "pillars" exist to provide dynamic stability:

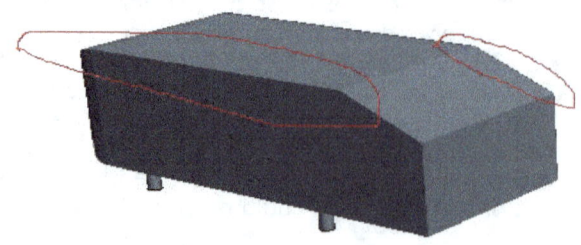

If there are edges that produce harmful turbulences downstream to some part, we can soften them to eliminate or at least mitigate its effect:

If the edge (red) is <u>parallel to wind direction</u>, is possible also generate a vortex pillar; why?

Because this edge, due to friction (viscosity), produces a pressure tube.

Depending if in zones blue or green there is fewer or more pressure, the vortex will rotate in one direction or another:

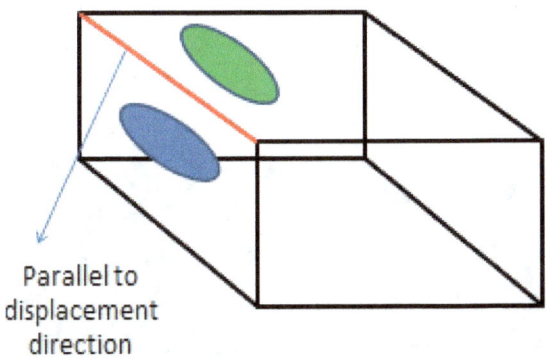

Parallel to
displacement
direction

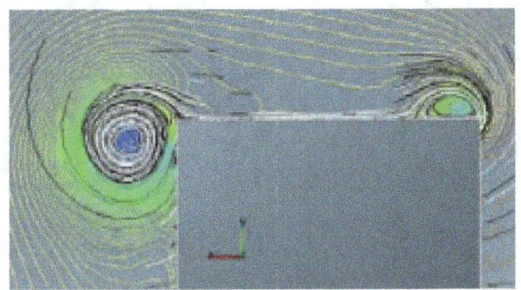

We can see these "pillars" in those corners which are not "rounded or smoothed" (which generate a lot of drag as in this case):

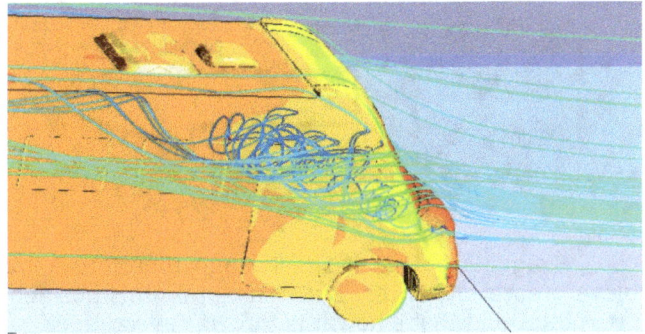

Vortices generation: another concept or idea.

We know that a bullet before leaving for the rifle barrel, rotates within the barrel. The goal is to provide rotation (inertia) to go "straight":

The bullet keeps rotating until it reaches its target.

The same happens when we throw a rugby ball: given its peculiar shape it must travel in this position with respect to the direction of the air.

Drag will be much smaller and therefore it may achieve greater distances. But to maintain this position it must rotate, and that's how it is thrown: in rotation:

Therefore, we can think of an alternative system, to generate high-energy rotational vortices, different to what we have seen up to now:

We can try to fold the airflow so that we keep the rotation applied previously. This could be done by circulating air through a kind of auger, so that we force a spiral rotation:

When air circulates this device entirely we think that it will continue to maintain the rotation a certain distance after; however, this is completely wrong: just at the end of the auger its rotation is drastically and suddenly interrupted. It is always necessary to generate vortices due to a device that is based on pressure differences; why ?

Because by this way, it produces a depression tube after system which suctions the air producing the vortex along the time:

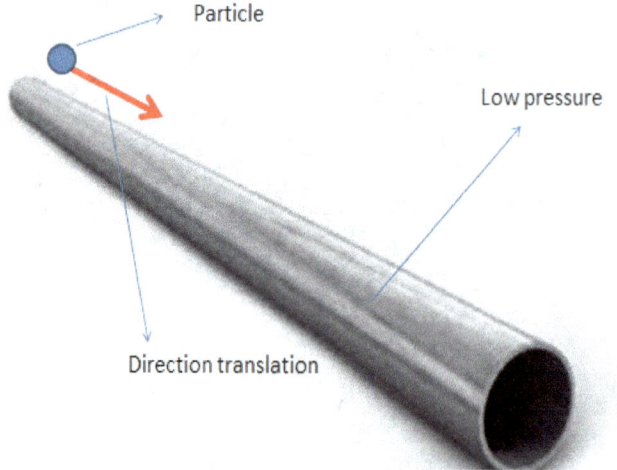

Particle

Low pressure

Direction translation

The translation of particle, produce the vortex.
➜ Important:

Let a spacecraft around planet; he have a speed and the gravity force "suction" the spacecraft to planet; depending the distance, the gravity force and speed, the orbit will be different (Newwton theory gravity):

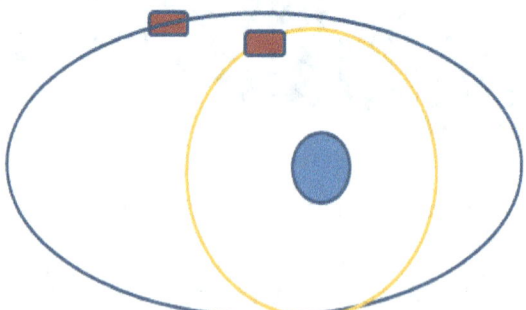

Analyzing these orbits around vortex center, it should be possible to know the vortex geometry, his force, his evolution, his combination, etc....

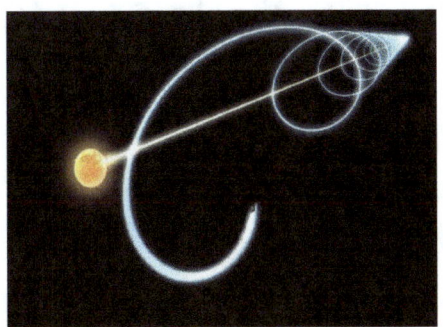

Some equations and relations about; "r" distance between air particle and vortex center, "v" speed particle, "P" pressure "attraction" center vortex, "G" constant vortex and "T" rotation periode:

$$circular - orbit \rightarrow G\frac{m}{r^2} = m\frac{v^2}{r}$$

$$T = 2\pi\sqrt{\frac{r^3}{Gm}}$$

$$P = \frac{G}{r^2}$$

Let

"m=1":

$$F = G\frac{m}{r^2} = \frac{G}{r^2}$$

Calculating the velocity rotation by CFD simulation, and trying that the same velocity is the result of Newton theory, is possible so, know "G".
That is very very important.

If let Force by Newton = Force Centripetal, we can know the orbit behavior. That is: what is the "G" value for our case

$$\frac{mMG}{r^2} = m\frac{v^2}{r}$$

$$\frac{MG}{r} = v^2$$

$$G = \frac{v^2 r}{M}$$

If "a" is the semi major exis

(orbit), amd "T" the periode:

$$a = \sqrt[3]{\frac{GM\,T^3}{4\pi^2}} \rightarrow \frac{a}{T} = \sqrt[3]{\frac{GM}{4\pi^2}}$$

$$G = \frac{a^3 4\pi^2}{M\,T^2}$$

Suppose us deflecting airflow through the deflector as follows:

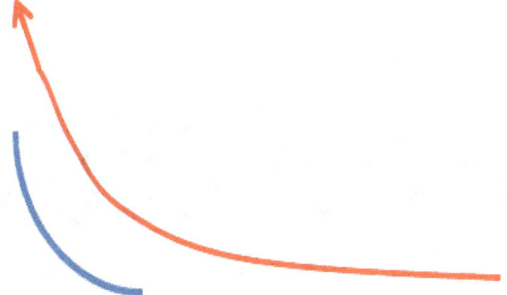

The direction marked on red, would be "ideal" or "desired"; but reality will different from this (loss energy); so is necessary to deflect air, that the air have energy (speed, pressure, etc):

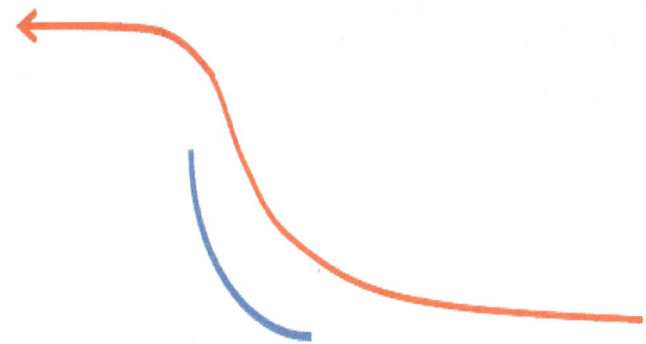

Once we stop applying a force to the flow, it ceases to modify its trajectory (direct application of the inertia principle ....). To modify the path of a flow, we must push it continuously and smoothly to achieve what we want to achieve. This is a fundamental principle.

Let's take a look at how pitch interferes with downforce:

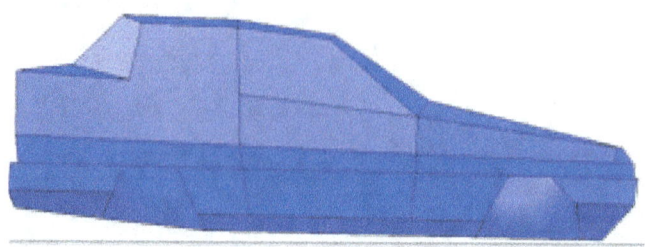

Suppose now that we vary the angle of attack with respect to ground making the height at the front smaller than the height at the back; this angle is called "pitch". This angle greatly varies the amount of downforce generated by the floor.

These height variations are usually very small in competition: the front and rear heights (car at full speed) are usually 4 mm and 7 mm only, or even less....

The pressure profile under the car has the following shape, at 0 ° of pitch:

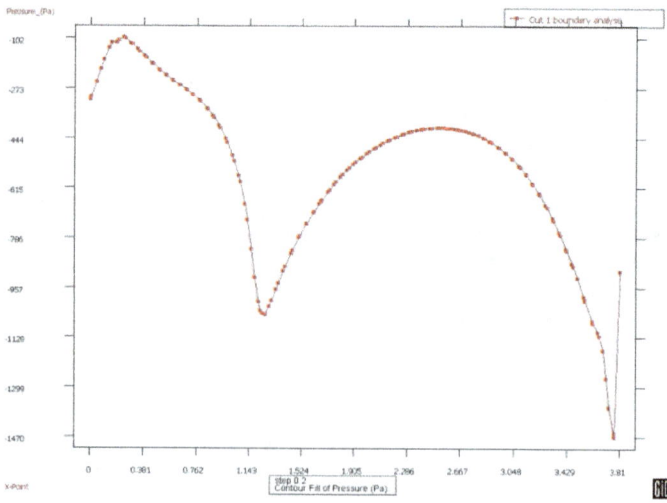

If we increase the pitch to1º, the graph becomes:

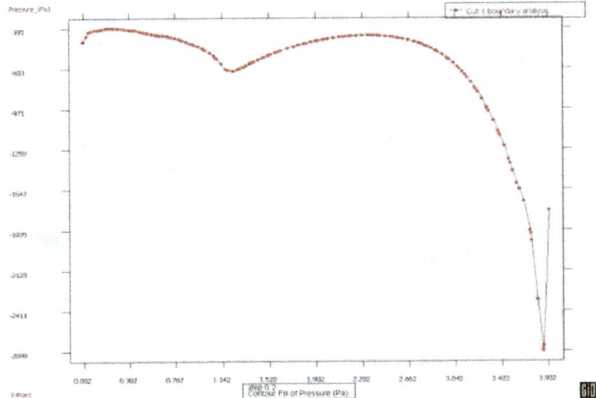

We appreciate that depression increases; i.e. more downforce is generated; but if we keep increasing pitch downforce stops increasing. If we recall the definition of a wing's stall angle, we can define similarly pitch stall. This limit cannot exceed 1º or 2º.

We see in the pictures below, that with 1º pitch, there is some flow which is being introduced below the floor spoiling the generation of downforce:

0 º pitch:

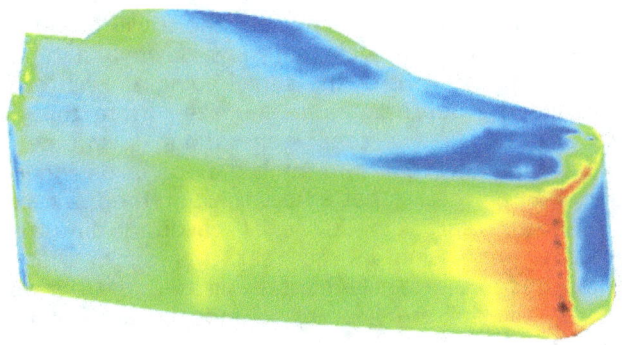

1ºpitch:

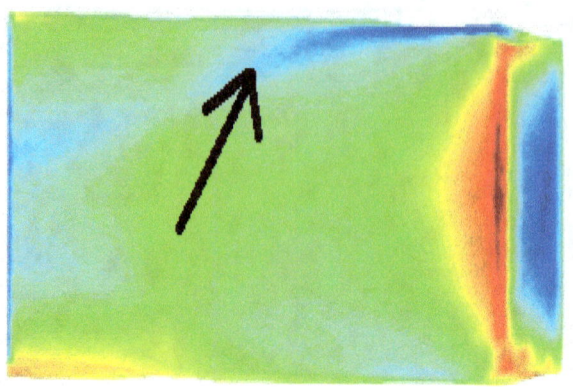

DIFFUSER

When trying to generate as much downforce as possible in a racing car's floor, on its own it doesn't generate all the downforce it could if the airflow doesn't travel at high speeds across the entire floor.

Suppose we have a rectangular floor and we channel high velocity air across it: there will be an area of the floor where the air will slow down and tend to leave and fill the floor or the surrounding air will tend to fill-change the depression; in any case, from this area onwards, the floor "stops generating downforce". To prevent this, the rear part of the floor is designed to ensure that all the air that enters through the front exists through the rear. This device is called a diffuser: it creates a depression in the rear that sucks the flow passing under the car; without a diffuser, the air would enter an area of the floor as it finds an area of lower pressure:

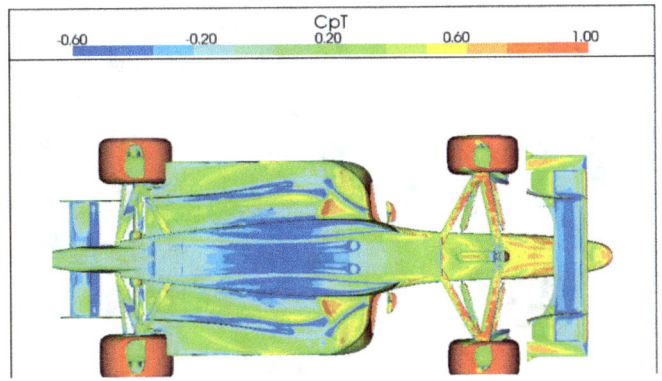

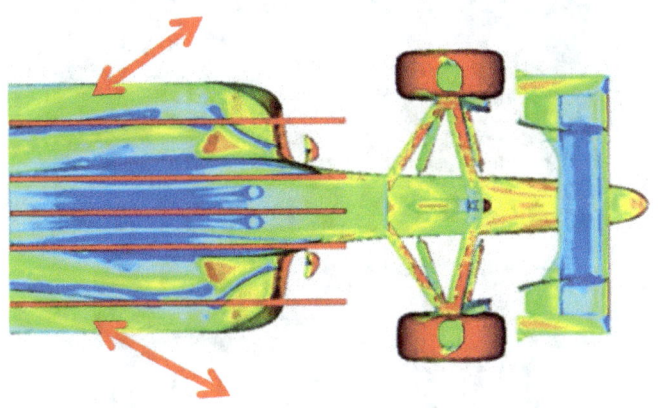

Low pressure:

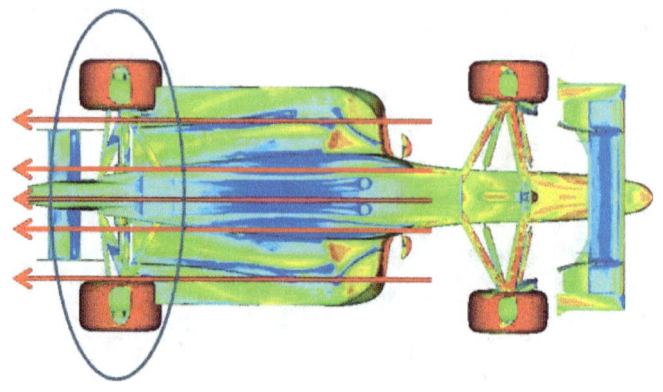

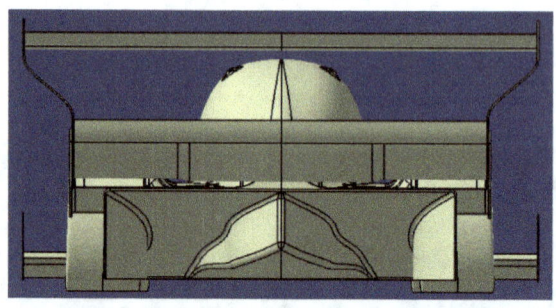

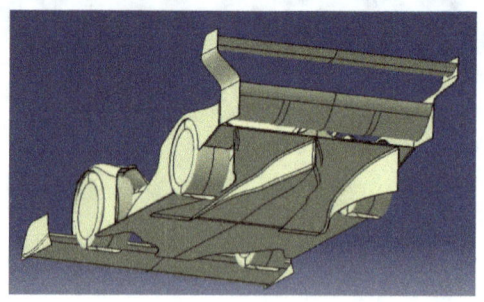

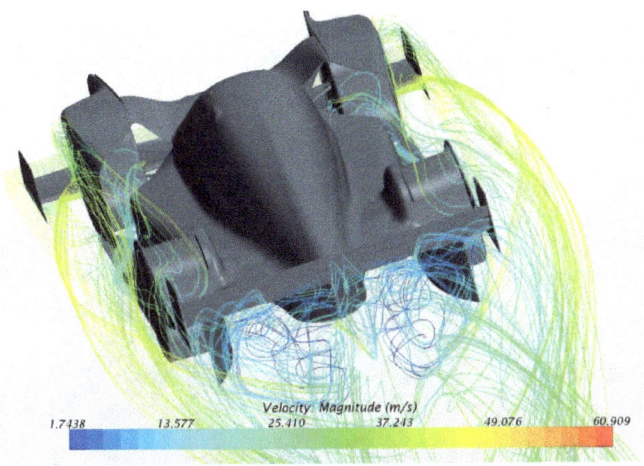

Velocity: Magnitude (m/s)

1.7438   13.577   25.410   37.243   49.076   60.909

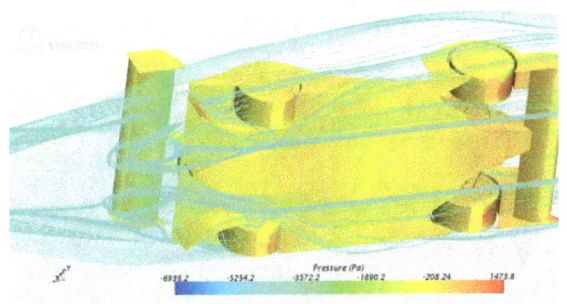

Pressure (Pa)

-6936.2   -5254.2   -3572.2   -1890.2   -208.24   1473.8

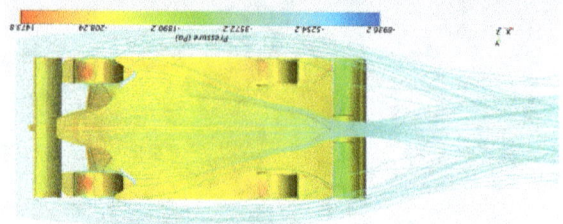

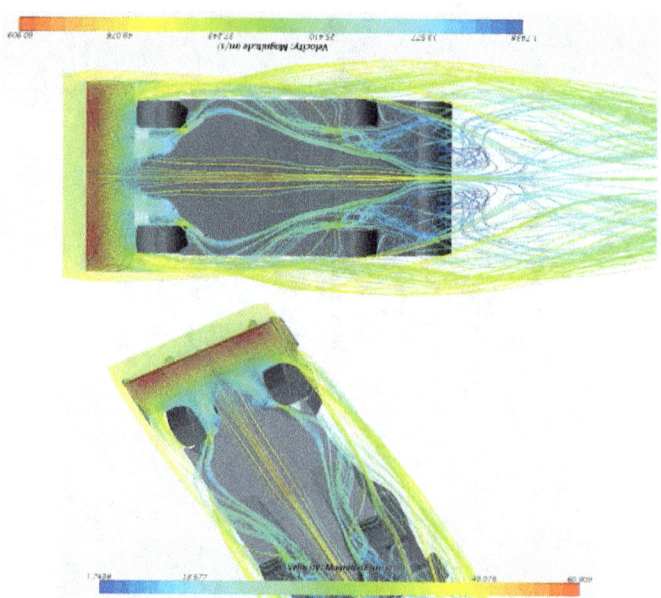

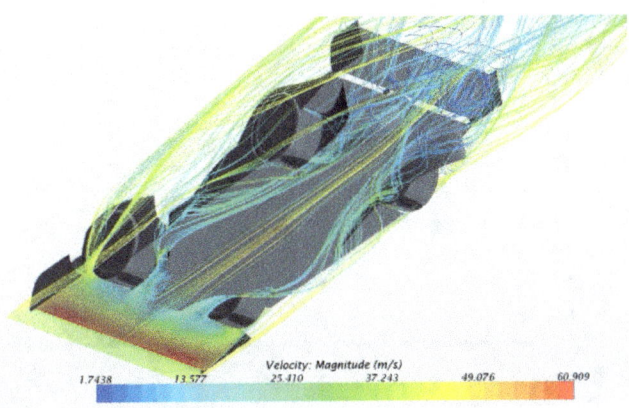

This "low pressure", is created in the named
"Crack line"; this line define the transition between floor
and difussor:

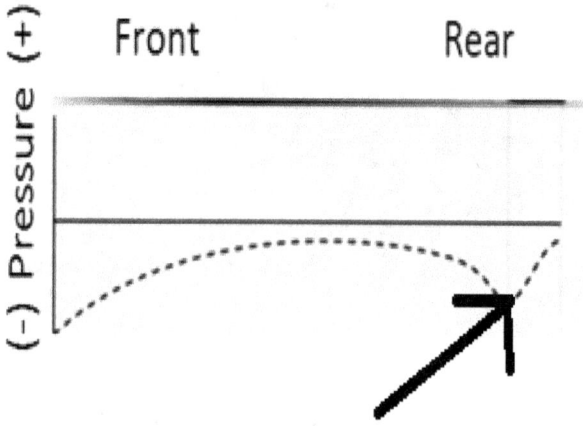

There two conservation principle involved there:

Conservation of mass.
Conservation of energy.

The complete assembly of undertray+diffuser form a tube of section S1 and S2 respectively.

The quantity of air that enters S1 per second must be the same that exits S2 per sec.
Since we're dealing with particle moving we can also say that the speed at which the quantity flows through S1 must be proportional to the speed that flows through S2 i.e $S1V1=S2V2$.

Since S2 is larger that S1, V1 is greater than V2 so the speed in the undertray is the greater than in the diffuser.

Now why does the diffuser accelerates the flows in the undertray?

This is because the air is slowed down in S2...but the density stays the same (at the speed a car travels); the word "expand" is misleading, the air is not stretching is actually just slows down...and thus a "vacuum" is created and since nature doesn't like it it will be filled...by air coming from the undertray...in other words the diffuser will pump air from the undertray to fill the whole S2 section and the S1V1=S2V2 principle will be verified, only that V1 will be even greater than if the diffuser wasn't there and S2 will be slower.

The peak acceleration at the location where the gradient of section is the greatest, since the difference of section will dictate the speed, the quickest and most abrupt change of section will lead the two steepest change in speed and this is where the diffuser starts.

Now, speed are defined, and they're changing, but energy's not. So basically the three (negating the internal energy of the system) forms of energies (derived from static pressure, height and speed) will varying to conserve the total energy of the system.

If speed increase, either potential energy (from height) or static pressure derived energy will need to compensate.

If we take one particle (in fluids dynamics a particle is not one molecule but a bunch of molecules, that is a small volume of air), its potential energy will not change (the particle will stay at the same height) so only the static pressure will decrease.

The static pressure will decrease and thus hopefully the pressure above the car will be higher thus creating a pressure differential and thus downforce.
The diffuser also generates downforce, because it has less pressure than the pressure "outside" the car; but the main goal of the diffuser is to make the floor work properly.

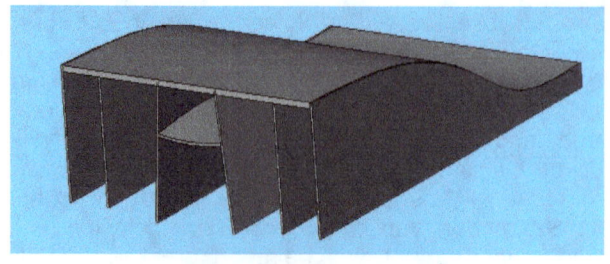

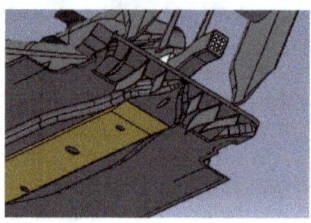

On the diffuser some dividers are placed, whose missions are:

- Divide all flow going downstream distributing the pressure across the entire

diffuser.
- Better management of this flow.
- Adjust the flow where it is needed or required.
- The "boxes" formed by these dividers produce depression, sucking the air from the floor.

It is convenient, if allow by the rules to "start channeling" the flow from as far forward as possible, so that there is less risk for the air to escape to the sides:

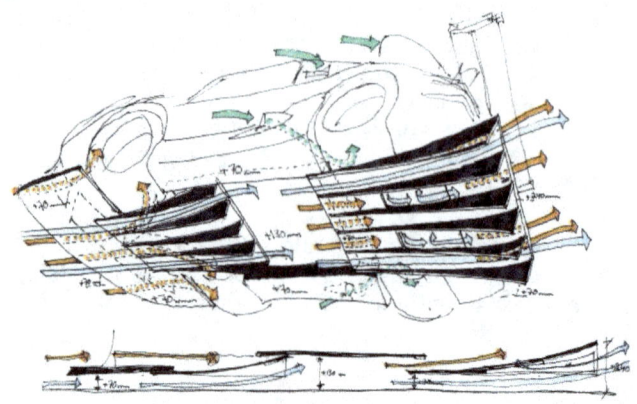

Another motive very important, is expand the air in axis "y"; we have seed that the expansion in "z" is essential; the expansion in "y" is not important, but is possible to allow some think about improve the downforce and filling the depression rear wheel:

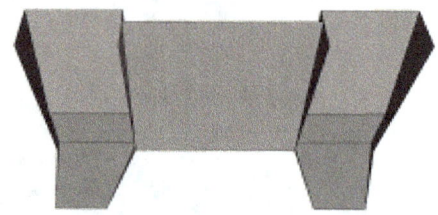

Is possible to think that the best diffuser-floor, is half trumpet: but not:

Difusser

There is a part which is often included to the rear wing, called diffuser spoiler or "wing beam"; this is a wing just above the diffuser and below the rear wing, which is responsible for generating a depression.

Its aim is to make the diffuser work properly; the famous Double Blown Diffuser, works the same way:

""**Why is necessary that the air arrive to the diffuser ? this is not because the diffuser generate a lot downforce: if the air reaches the diffuser, means that the air has travelled all the ground and the ground, is the principal element about downforce.**""

If the rules allow it, we could give very different shapes to floors and diffuser in a more efficient way than the traditional flat configuration:

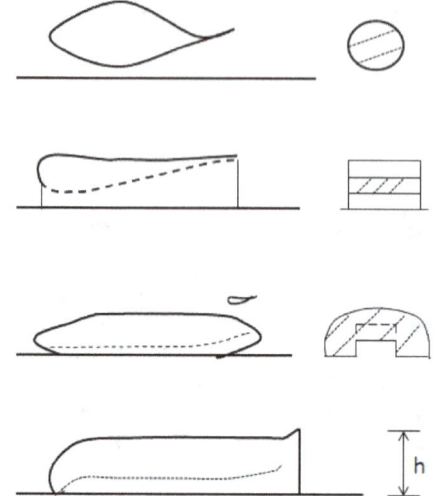

Also with an extractor fan:

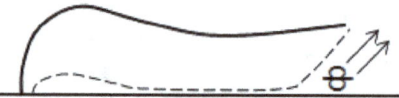

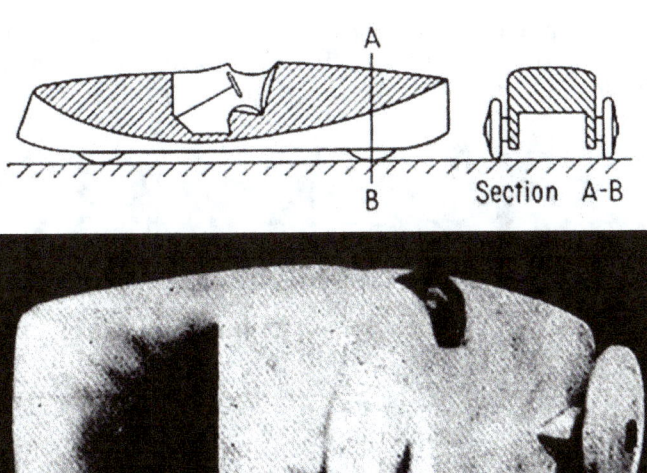

Section A-B

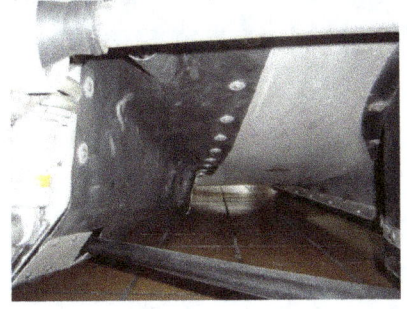

In fact, Old people are older, but not stupid; ground effect is old; Aerodynamics is an "old" science. Wisdom is even older. Age means nothing:

The exit angle or diffuser angle is very important; Let's take at look at its relevance plotting the drag and lift coefficient as a function of the diffuser angle:

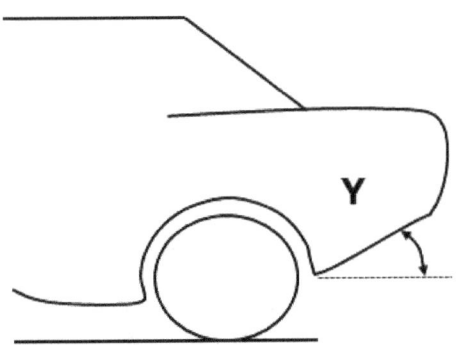

Rear Downforce: we must know this variation as a function of the diffuser angle to optimize the car:

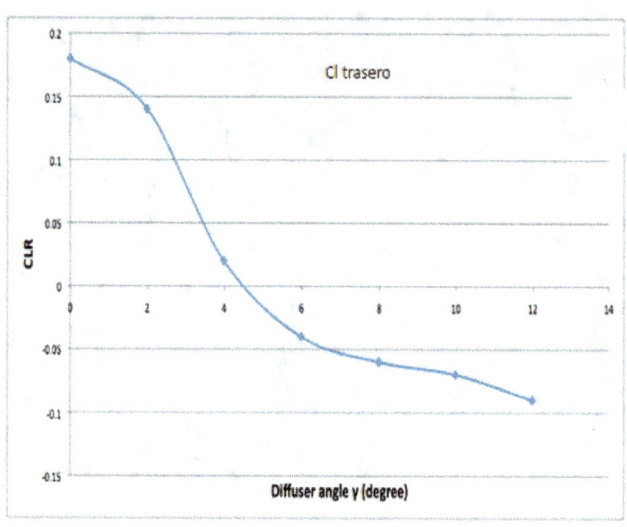

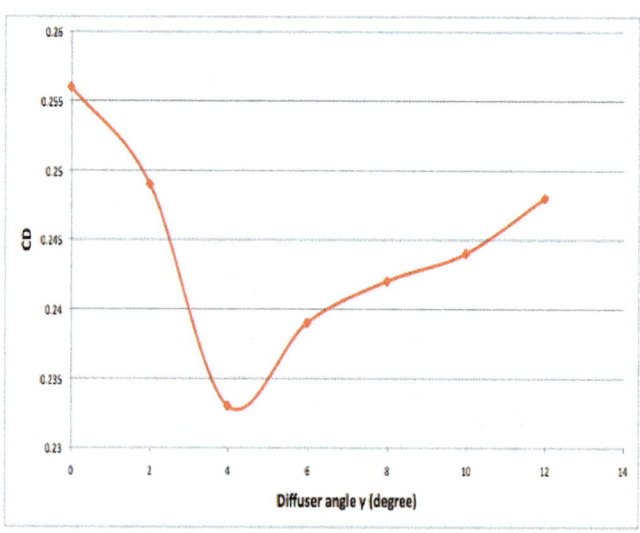

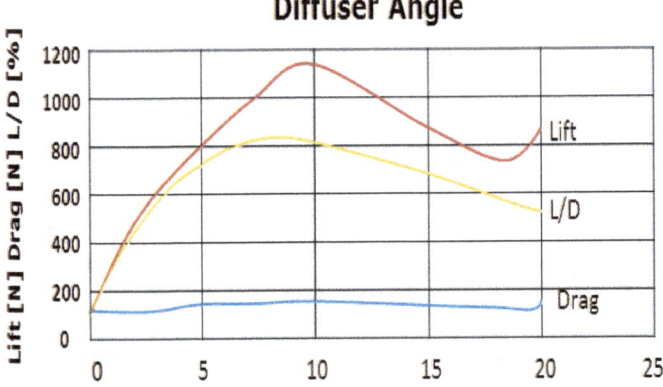

Diffuser Angle

We must also know the variations of the lift and drag coefficients as a function of the ride height:

We work on a diffuser, measuring downforce and drag and efficiency "L / D":

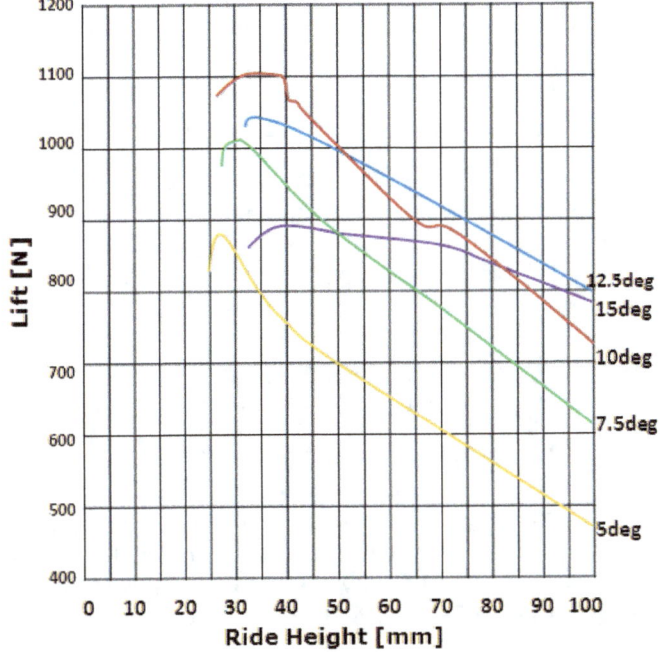

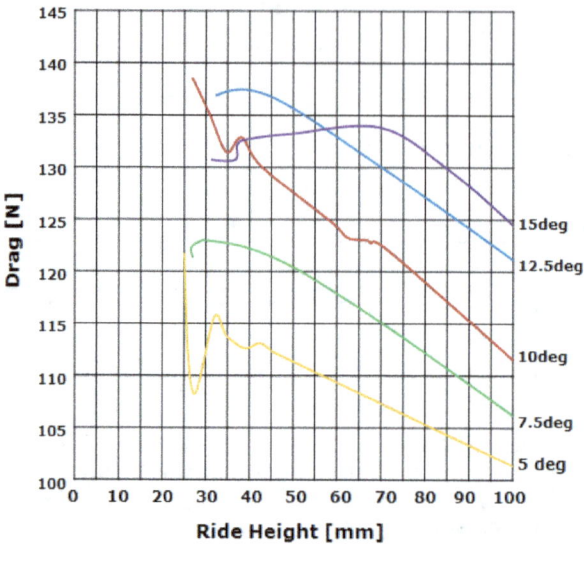

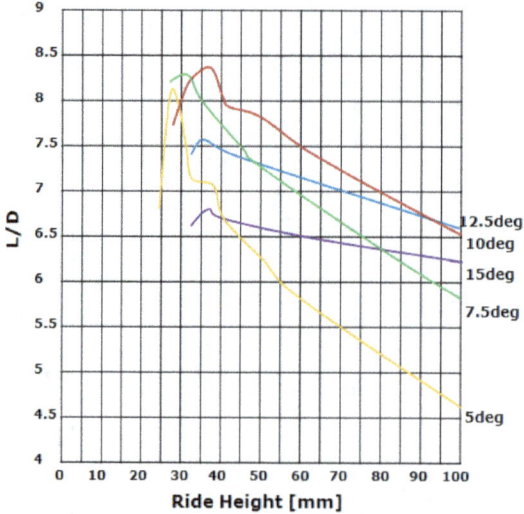

With this study we can determine the optimum angle for the diffuser and the ride height of the vehicle, in this case it is 37mm. obviously this height is variable but the diffuser angle is fixed.

In addition to its shape, another fundamental parameters that matters to the amount of downforce generated by the floor, is the angle between the track and the floor and distance to the track. It is very important to get a good setup of the car, study and understand these variations:

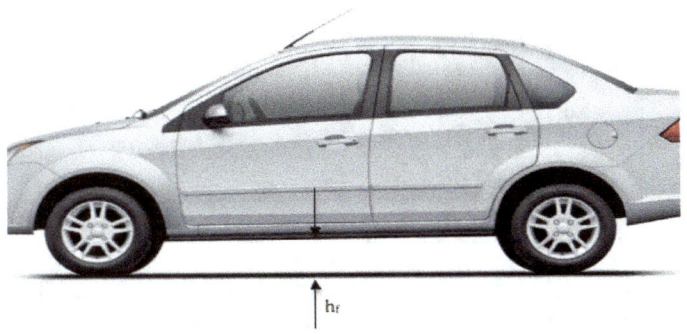

h_f

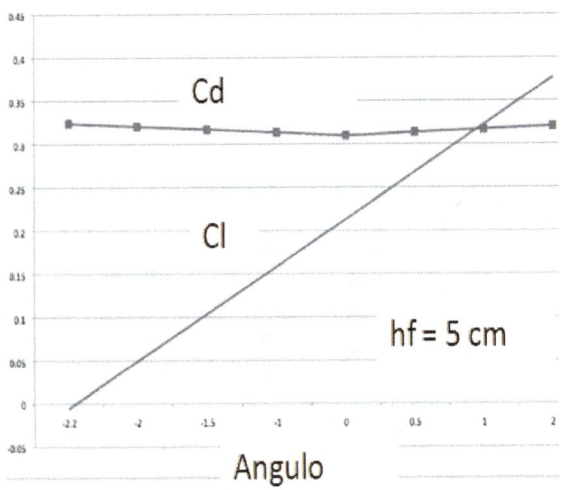

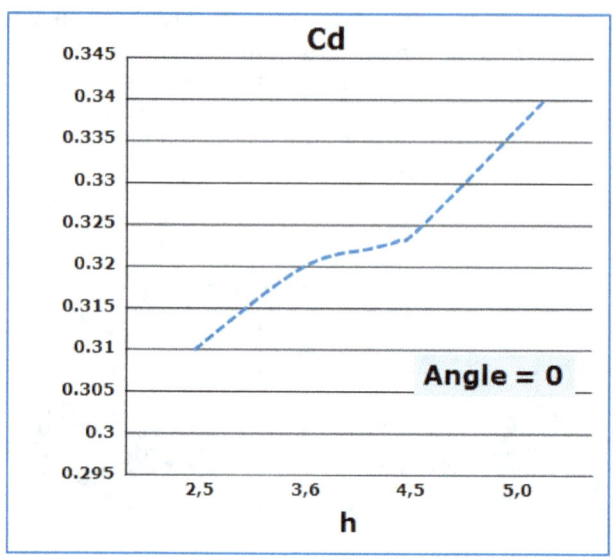

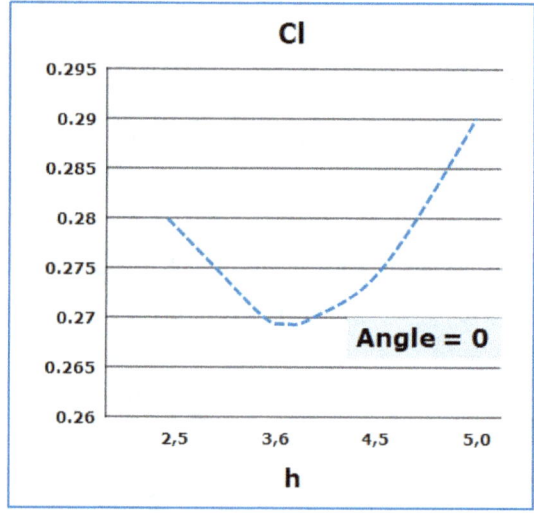

In the case of a racing car pressure variations at the bottom of the car are more prominent and sensitive:

Depending on the heights relative to the asphalt (h), we can see that the pressure under the car (position "x") reaches negative values increasing if the distance is smaller; distances close to 0mm as shown previously are impossible:

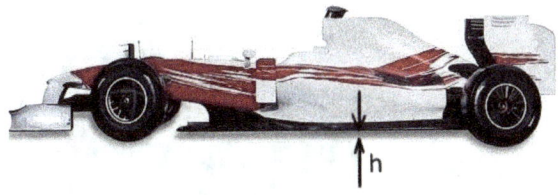

Let the take a look at the variations in the pressure generated depending on the height "h":

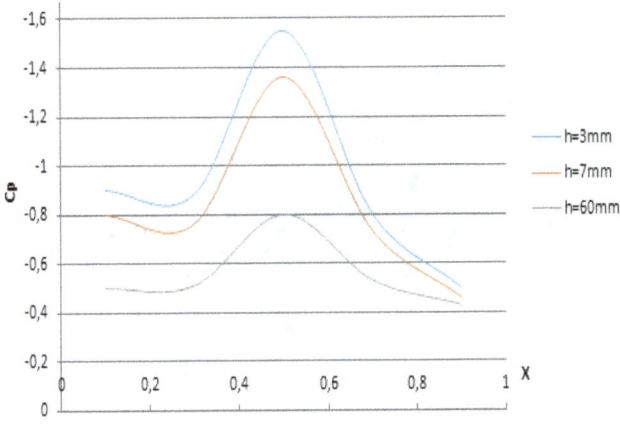

If the diffuser stalls, i.e. the angle of the diffuser is too large and the flow detaches from the diffuser, it can have serious consequences to the aerodynamic performance of the car because the floor stops working as downforce generating downforce device. The importance of the diffuser is huge, whichever way you look at it: remember the importance of the floor and diffuser to the total downforce generated.... We can use CFD to see if the diffuser stalls and where, "seeing" abrupt changes in color or pressure:

CpT
-0.20   0.20   0.60   1.

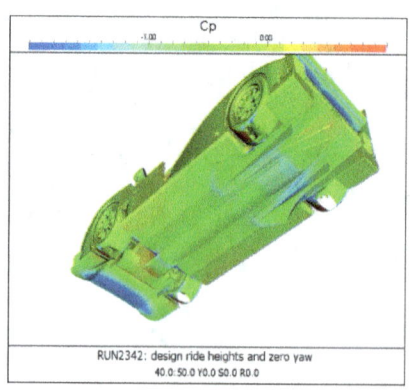

Cp
-1.00   0.00

RUN2342: design ride heights and zero yaw
40.0: 50.0 Y0.0 S0.0 R0.0

To further increase downforce, we can vary not only the speed of the air moving through the floor, but also the amount; this would also affect the amount of downforce:

Another function of this "hole frontal", is allow air always through the ground....

Summary:
How work so, the difusser in a race car ?

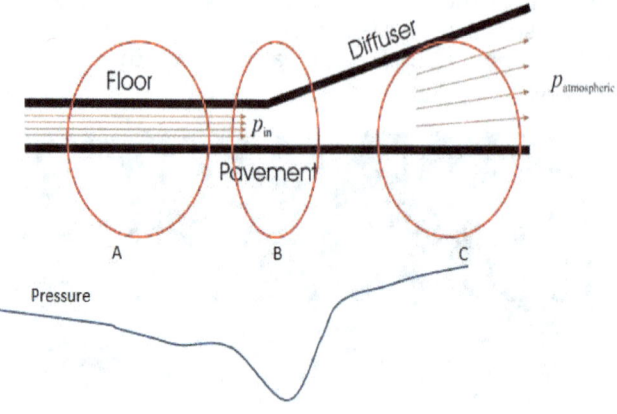

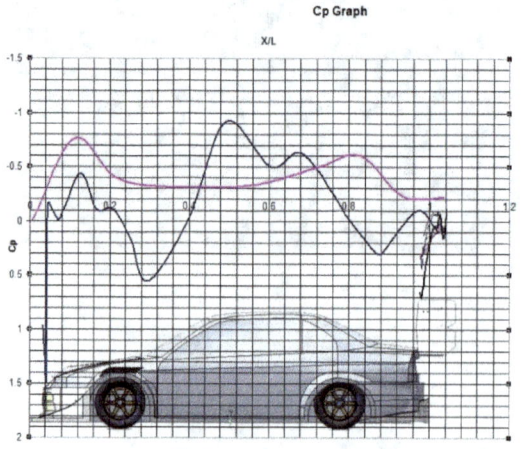

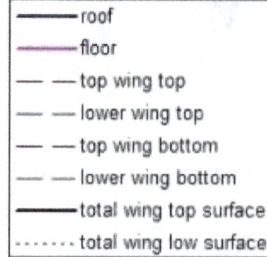

By Bernouilli, in A, the speed is high so the pressure is low (create downforce); the problem is that the air outdoors, want to entry in A, losing the downforce; is necessary study the amount air and speed air (both together):

In C, by Bernouilli too, the pressure is higher than pressure in A; how so, is possible create downforce in C ? because the pressure in C, is higher than atmospheric pressure.

The most important actions of the diffuser, are two:

- Create downforce in C, even though the pressure in C is higher than A; so, how is possible that the diffusor help to ground to suck the air ? because:
- In transition A to C, exist a low low pressure (crack pressure); this low pressure sucks the air in A.

That is the basic operation of diffuser: very important to creating downforce, because help to ground.

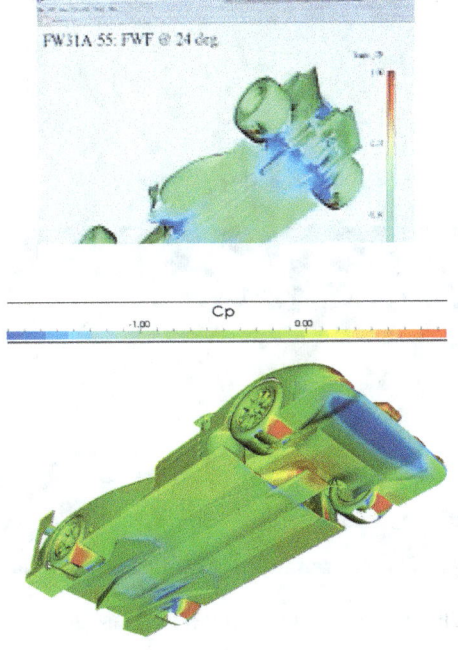

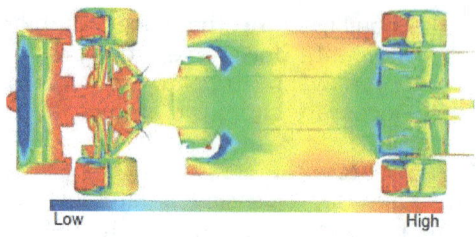

Low   High

We can see the same "crack pressure" in the diffuser of F1:

We can see the same "crack pressure" in the diffuser of F3:

We have the next floor + difusser, for analyzing that:

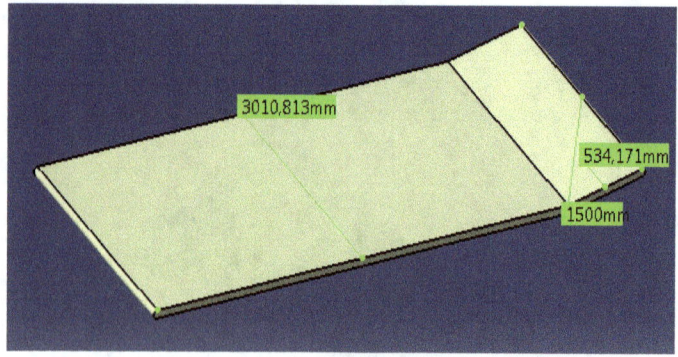

Without boundary layer; pressure distruibution surface below (floor and difusser):

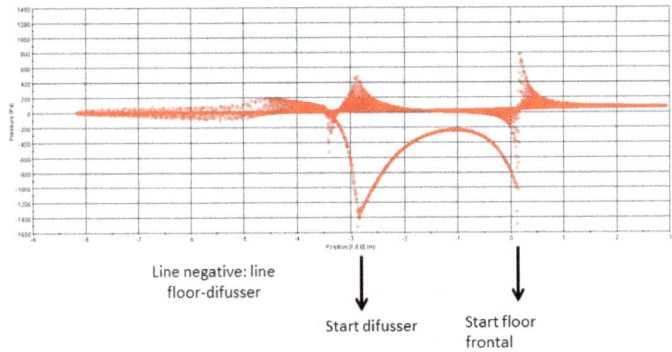

Line negative: line
floor-difusser

↓   ↓

Start difusser   Start floor
                     frontal

With boundary layer; pressure distruibution surface below (floor and difusser):

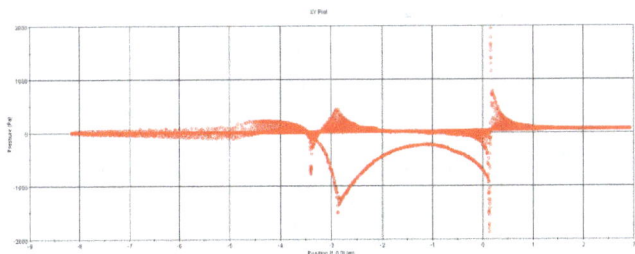

All pressure below floor and diffuser is negative (downforce); that is the important iiii.

We can think about one nozzle, in order to understand the "crack pressure":

The "crack" is the same phenomenon; that is produced by area change:

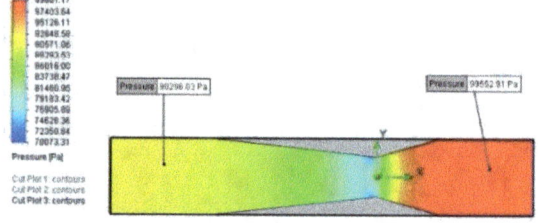

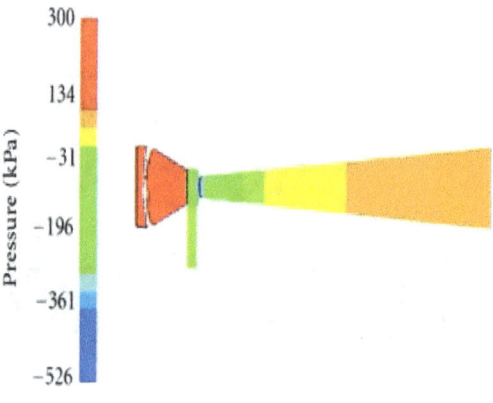

SPLITTER

This is a forward extension of the floor; it generates an increase of the front downforce and a small reduction of the rear downforce:

Front: + 40% L
Rear: - 7.9% L
Lift / Drag: + 5.9 %

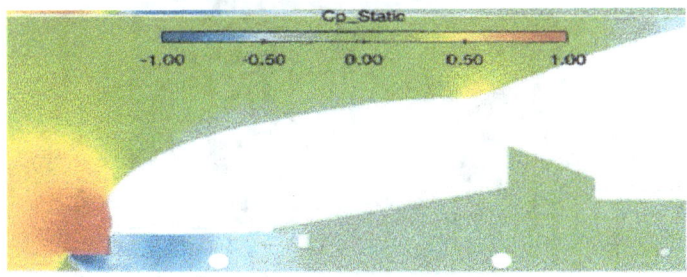

Consider the influence on drag and downforce:

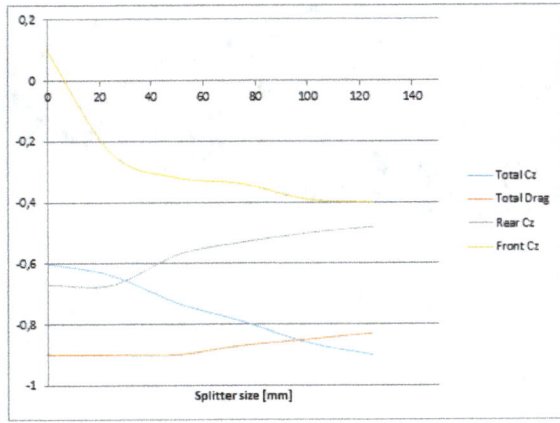

AIR DAMM

It is a screen or frontal barrier, which prevents the airflow entering "normally":

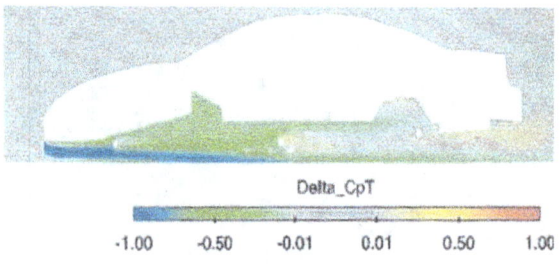

Let's plot variation of the aerodynamic parameters as a function of the height or size of the Dam:

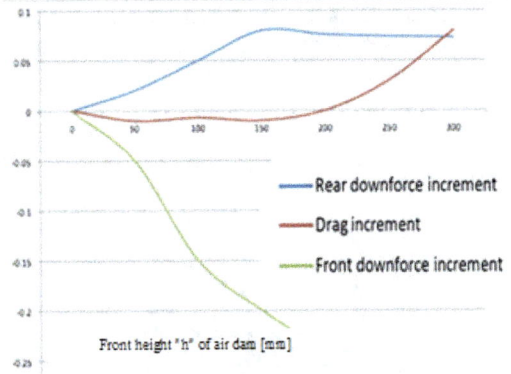

- Rear downforce increment
- Drag increment
- Front downforce increment

Front height "h" of air dam [mm]

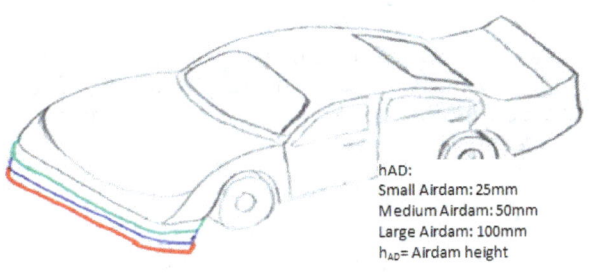

hAD:
Small Airdam: 25mm
Medium Airdam: 50mm
Large Airdam: 100mm
h_{AD} = Airdam height

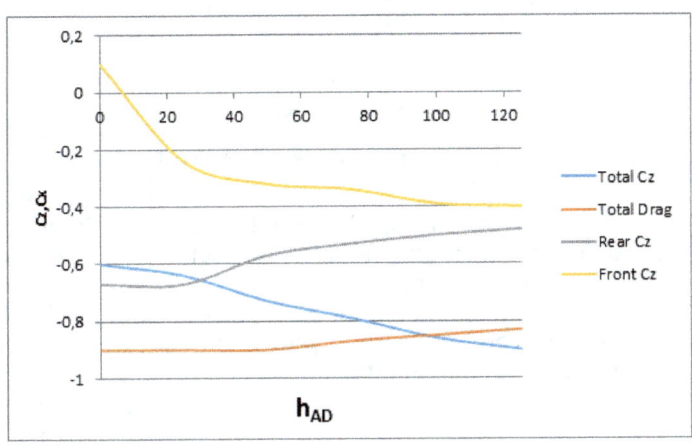

- Total Cz
- Total Drag
- Rear Cz
- Front Cz

h_{AD}

This device has some disadvantages, such as the need of complex brake cooling systems or oil radiator, as the device "prevents" the air travelling across; you need to add a few "extras" openings through which cooling air can enter:

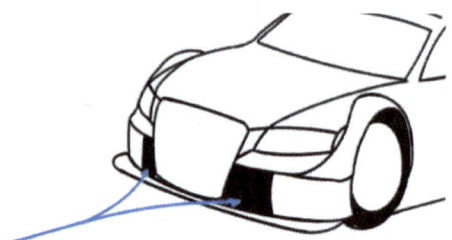

The purpose of an airdam, is to generate a depression in the lower front area.

The air dam sometimes can help cool the engine since we can place an opening to extract of hot air and use the depression generated in the lower area to help such extraction:

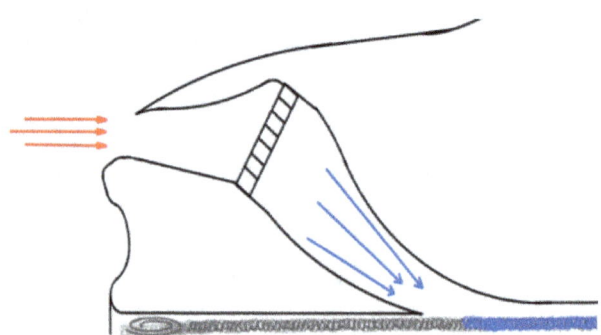

In many cases, the Splitter and Air Dam go together and work together; it is useful to know the interaction between the two.

Suppose the study of a car with an airdam of 100 mm; we will compare results with a different case where we include a 150mm splitter:

Velocity 50m/s   Area = 1.3m^2

Element	100m Airdam	100mm Airdam + 150mm Splitter	Drag Increment
Bodywork	569 N	577 N	-19N
Wheels	213 N	196 N	-18 N
Succion	194 N	238 N	+44 N
Accessories	56 N	111 N	+55 N
Wing	267 N	266 N	-1 N
AirDam	337 N	317 N	-20 N
Splitter	0	0	0
Total	1663 N	1705 N	+42 N

There is an overall increase in the drag of 2.5%, while there is also an increase in 10% downforce.

FRONT DIFFUSER

It is installed only those racing categories where rules allow it; in LMP2 and LMP3 cars the floor's front area is divided in two: at the front you can place this system; the rules clearly state that the plate that forms the "diffuser" MUST HAVE NO PROFILE or section variations.

This "front diffuser" is a divergent duct that doesn't cause stall similar to what happens with the rear diffuser.

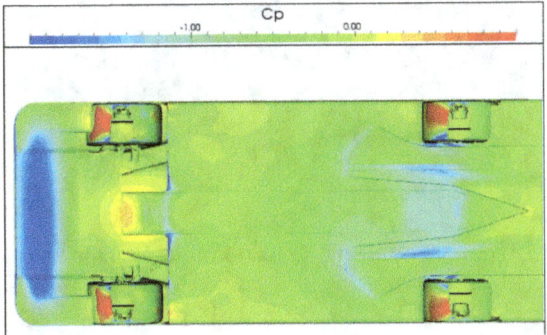

The airdam + splitter + front diffuser solution provides an "extra" downforce, particularly applicable to the Nascar series:

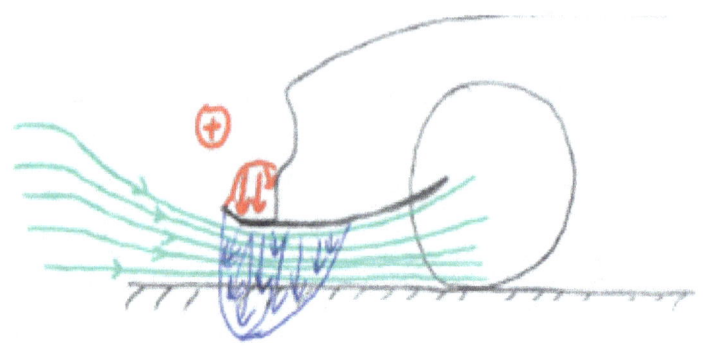

The aerodynamic load gains in Nascar are:
$$\Delta C_z = + 3.9 \% ---- 4.2 \%$$
$$\Delta C_x = + 1.4 \%$$

Gaining a 2.56 % in efficiency (L/D):

$$\left[\frac{\Delta C_z}{\Delta C_x}\right] = \frac{1.04}{1.014} = 1.0256$$

Note:

Any vehicle moving on a "ground", tends to produce an air mattress at the bottom; aircrafts and other vehicles use this to keep "flying" with little energy; so does the hovercraft; in the case of racing cars this is different, because we want to generate the lesser pressure under the car as possible:

STREET VEHICLES In street cars, ground effect and ride height work differently.

Suppose "h" the distance between the car's floor and the road and "b" the wheelbase; thus, we can observe that the larger ground clearance, less downforce is being generated, but higher drag:

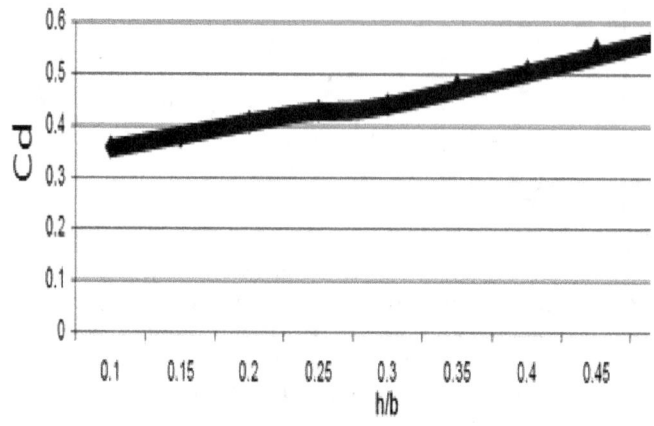

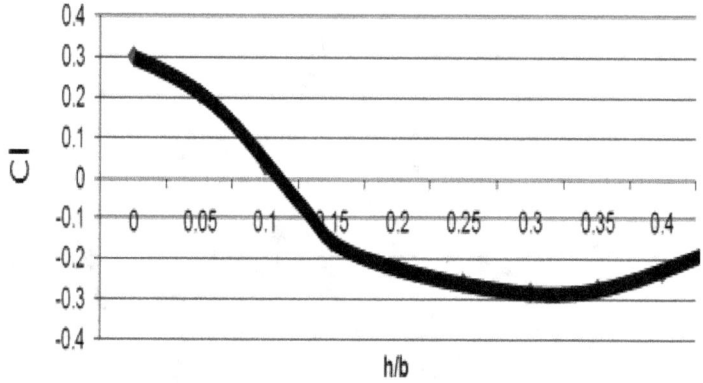

SPLITTER OPENING

We already know the function of a splitter which is to increase downforce on the front axle.

The splitter reduces the amount of air passing under the floor, so that the rear axle is affected, reducing the downforce generated in that area. You may sometimes need less front downforce, and we can produce excessive oversteer, besides increasing resistance; therefore, we can make the splitter smaller.

On the other hand, you may also need more load at the rear or just not advance in excess the pressure center; to achieve this effect, we can make an opening to the splitter to:

- Reduce the front load.
- Increase rear loading.
- Locate further at the back the pressure center.

In fact, if we were to reduce a lot the front opening or the distance to the ground, the diffuser would not have air implying a drastic reduction in downforce (airflow is required below the floor):

Splitter with central opening:

If augment the air inlet from front floor, the thoughness to separation flow in diffuser will be lower (the boundary layer thickness increase in floor); that is very important to have a good design iiii

If the is braking, is possible reduce a lot the front height; that is: there is not front downforce (without opening splitter); for that, is necessary open it.

AQUAPLANING

One of the most influential factor that determines the heights of the car's floor is aquaplaning; this height will largely determine how the car will behave on the track.

Assume that "L" is the wheelbase of the car, "P" the tyre inflation pressure (KPa) and "W" the car's weight; We can define a simple relationship between the parameters described, to calculate the speed at which the tyre enter aquaplaning:

$$V = 48.7 \sqrt{\dfrac{P}{\left(\dfrac{W}{L}\right)}} \qquad \text{[km/h]}$$

Tyre Pressure [kg/cm^2]	Tyre Pressure [kPa]	V water (km/h)
2.5	245	99
2	196	88
1.7	166	81
1.2	117	68

The evacuation flow may vary between 6 L/s in the case of a street car, and 26 L/s in the case of a F1 car.

These speed values from which hydroplaning occurs, are essential to know which are the top speeds that we can reach. However, what also interests us from the aerodynamic point of view, is the ride height; remember that the car's floor is very near the track; This can occur even if there is little water on the track, aquaplaning.

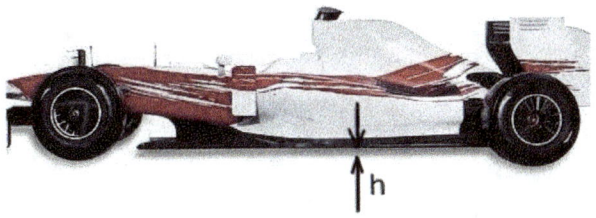

To avoid this important fact we must increase the height; when wet, ground clearance of a Formula 1 is approximately 15 mm higher than normal.

Another thing that we can do, if possible, is to increase the tyres diameter: larger diameter lower rotational speed; in Formula 1 tyres are about 10 mm larger.

Let's take a look at the aquaplaning while cornering and in a straight:

- In a corner:

Suppose a corner of radius 100 m and a water film of 7 mm; we gradually increase the speed of the car in the corner and we compare it with the ideal trajectory:

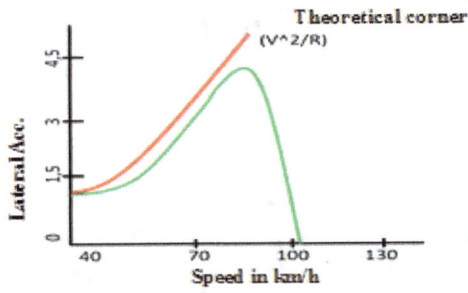

- On a straight:

Suppose 2 options:

    a) One part of the car is on a dry area and the other on a water surface 7mm.

    b) The entire car is on a water surface of 7 mm.

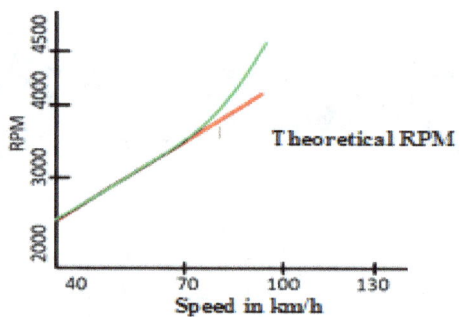

When the line begins to separate from the ideal, hydroplaning occurs.

If the vehicle had no sidepods and we could design a flat floor to generate lots of downforce, we could use the following concept design:

FLOOR PRESSURE IN CORNER

The distribution pressure in floor, when the car is in corner is something very important so this distribution or his variation has influence in full dynamic. Is necessary so, design the suspension and aerodynamic in function of this pressure field.

For example, for a 45 m/s speed and a 120 m / 25 m radius corner (pink and blue low pressure and red and green high pressure), are:

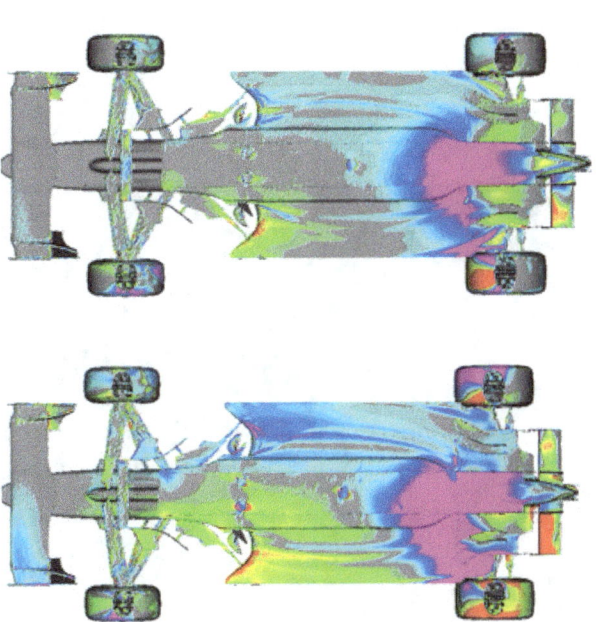

Obviously, this distribution pressure depends of:

- Height front and rear and his variation.
- Aerodynamic load and his variation.
    → So the suspension together aerodynamic loads.

Another examples in a F1 car: yaw = 0.5º (the path is different side by side):

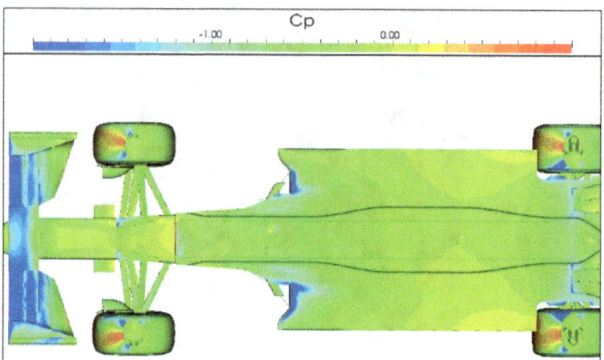

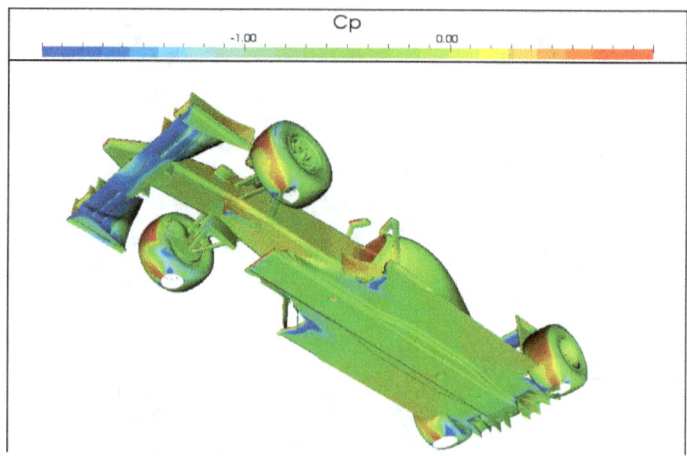

Now, in a LMP1:

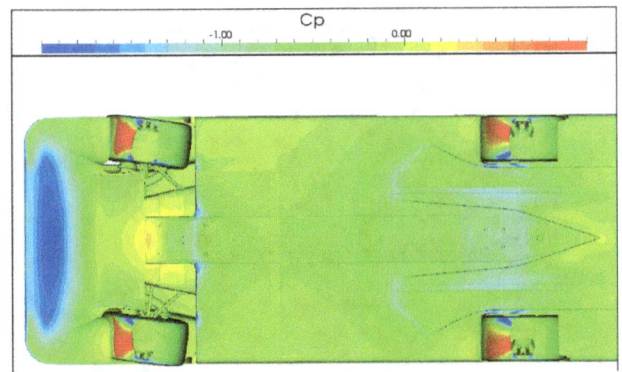

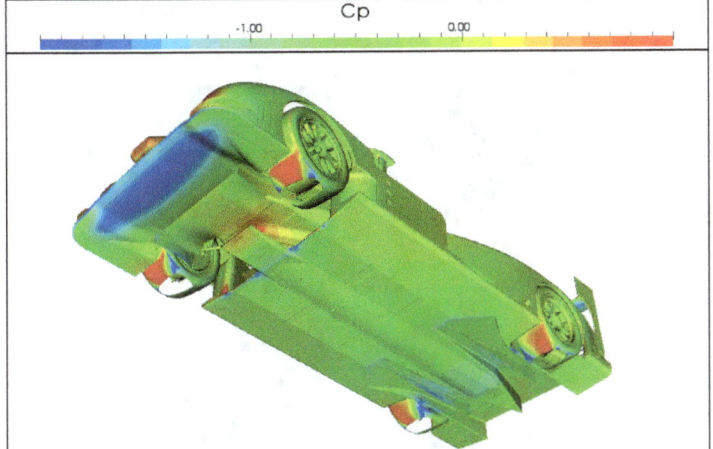

➔ Example:

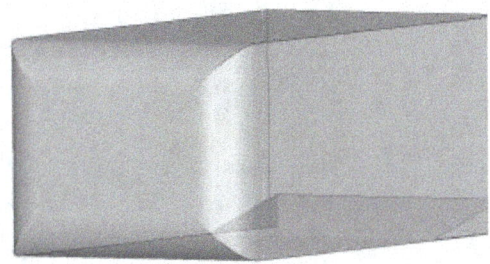

(a) front view

(b) rear view

In the diffuser, we see vortex creation:

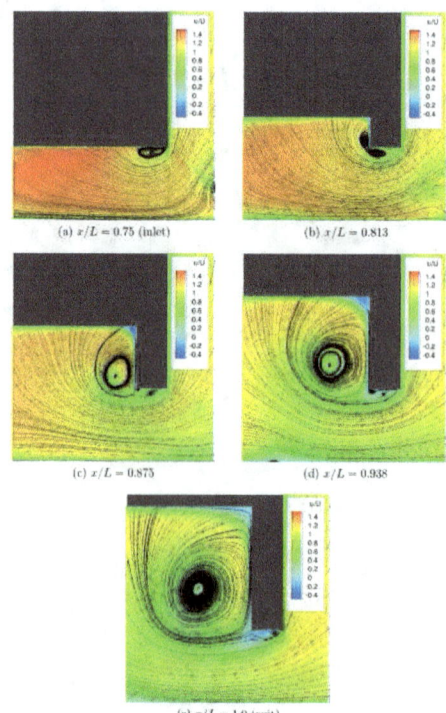

(a) $x/L = 0.75$ (inlet)

(b) $x/L = 0.813$

(c) $x/L = 0.875$

(d) $x/L = 0.938$

(e) $x/L = 1.0$ (exit)

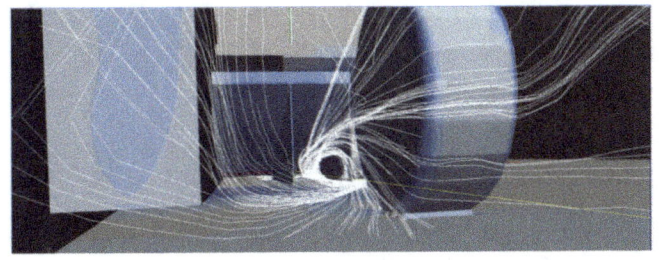

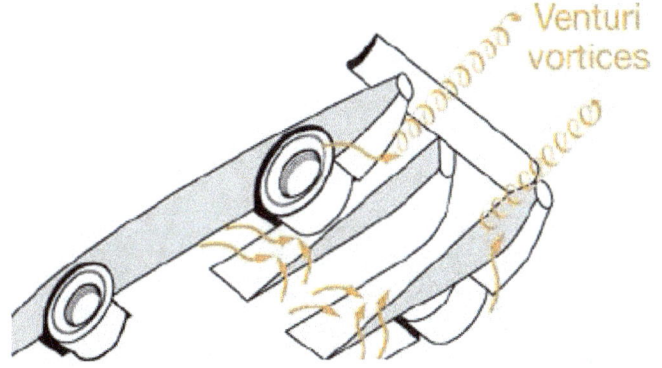

Venturi vortices

Resume:
THE VORTEX HAVE 4 CAPABILITIES VERY IMPORTANTS:

1. PRODUCE STABILITY.
2. PRODUCE LOW PRESSURE CONSERVATION
3. IS A SEAL
4. HELP FLOW TO GO YOU WANT
5. TO WRAP THE AIR (MORE ENERGY)

www.ingramcontent.com/pod-product-compliance
Lightning Source LLC
Chambersburg PA
CBHW071026220526
45467CB00004B/1527